Advance Praise f

SVO

Even as renewable biofuels come under s
with flying colors. Backed with documented research and hands-on experience, Forest Gregg leads us through the complexities of vegetable oils and diesel engines, coming out t
relatively simple means to run die
commercial cooking oil. *SVO* will do
what the many biodiesel books hav
and byproduc

— Richard Freudenberger, Publisher of BackHome Magazine

Forest Gregg's *SVO* is a definitive work. It's well researched, thorough, and will be an enduring reference point for the SVO community.

— Lyle Estill, author of *Biodiesel Power: The Power, Passion and Politics of the Next Renewable Fuel* and *Small is Possible: Life in a Local Economy*

Do you want to know how to unplug from petroleum and pollution and plug into community-driven, renewable fuel? This book is a must-read if you've got a commitment to finding independent, gasoline-free fuel sources and a penchant for DIY.

— Sarah Rich, managing editor of *Worldchanging: A User's Guide for the 21st Century*, and editor of *Dwell* magazine

SVO

Powering Your Vehicle with Straight Vegetable Oil

Forest Gregg

NEW SOCIETY PUBLISHERS

To Chris and Halle

Cataloging in Publication Data:
A catalog record for this publication is available from the National Library of Canada.

Cover design by Diane McIntosh.
Images: iStock/Benjamin Goode; iStock/P. Wei

Printed in Canada. First printing May 2008.

Paperback ISBN: 978-0-86571-612-4

Inquiries regarding requests to reprint all or part of *SVO* should be addressed to New Society Publishers at the address below.

To order directly from the publishers, please call toll-free (North America) 1-800-567-6772, or order online at: newsociety.com

Any other inquiries can be directed by mail to:

New Society Publishers
P.O. Box 189, Gabriola Island, BC V0R 1X0, Canada
(250) 247-9737

New Society Publishers' mission is to publish books that contribute in fundamental ways to building an ecologically sustainable and just society, and to do so with the least possible impact on the environment, in a manner that models this vision. We are committed to doing this not just through education, but through action. This book is one step toward ending global deforestation and climate change. It is printed on Forest Stewardship Council-certified acid-free paper that is **100% post-consumer recycled** (100% old growth forest-free), processed chlorine free, and printed with vegetable-based, low-VOC inks, with covers produced using FSC-certified stock. Additionally, New Society purchases carbon offsets based on an annual audit, operating with a carbon-neutral footprint. For further information, or to browse our full list of books and purchase securely, visit our website at: newsociety.com

NEW SOCIETY PUBLISHERS newsociety.com

Contents

Acknowledgments

I'd like to thank Joe Beatty and James Hudson for the many hours of invaluable conversation about vegetable oil, diesel engines, and their intersection, Anna Bogle for her enormous hospitality, and Michael Castelle for being an early reader and trustworthy adviser.

Using this Book

This is a book about using vegetable oil in a diesel engine, written for two different kinds of readers. The first group includes those considering using vegetable oil as an alternative fuel, and who need information that will allow them to decide if it is a good option — and if so, to make an informed choice about what vegetable oil conversion kit to purchase or what requirements are necessary to build an adequate conversion system themselves. The second group includes those who are interested in a deeper understanding about how vegetable oil interacts with diesel engines.

While this book can be read straight through, depending upon what your goals are, you may want to use this book differently. If you are in the first group trying to figure out if vegetable oil is right for you, the sections on System Design and Practicalities, plus the Guide to Diesel Vehicles in Appendix C will probably be of the greatest use. If you are looking for a deeper understanding, the chapters on Diesel Engines and the Fuel Properties of Vegetable Oil will be of the highest interest.

Beneficial Skepticism

This book represents the current state of our understanding of how vegetable oil fuel interacts with a diesel engine. As such, it is our expectation or at least our sincere hope that much of this information will be outdated within a few years. This is a rapidly moving field, and this book should be treated as a snapshot of where we are at, not as a fixed and forever thing.

Introduction: Is Vegetable Oil Right for Me?

In this chapter, we will lay out the argument for using vegetable oil as an alternative diesel fuel. We should note, first of all, that vegetable oil can only be used in diesel engines. If you don't have a diesel engine, then you also need to be convinced that it is worthwhile to invest in a diesel, in addition to being convinced that vegetable oil is a worthy choice.

There seem to be four different factors that motivate people to use vegetable oil as a fuel: environmental concerns, cheaper fuel costs, geopolitical concerns, and an interest in tinkering.

Environmental Reasons

The use of straight vegetable oil reduces the carbon contribution of a diesel engine significantly or completely, depending upon whether the oil is fresh or used. The tailpipe emissions of vegetable oil are probably comparable to that of biodiesel, which means that it is better on many measures and worse on some.

Fossil Fuel Efficiency

There has been no thorough life-cycle energy and carbon analysis of vegetable oil fuel, but we can make some extrapolations from work that had been done on biodiesel. In a 1998 study, the National Renewable Energy Lab looked at how much energy from fossil fuels it took to get biodiesel to the fuel tank, including how much energy it took to grow the crop, transport the crop, crush the crop into oil, transport the oil, convert the oil into biodiesel, and transport the biodiesel.[1] They found that every megajoule (MJ) of biodiesel at the pump took .311 MJ of fossil fuel energy to get there. This gives a fossil fuel efficiency of 3.215. In other words, by producing biodiesel we can turn one MJ of fossil fuel energy into 3.215 MJ of energy.

The fossil fuel efficiency of straight vegetable oil should be better than that, since, unlike biodiesel, it does not require fossil fuel inputs as raw material, or require an energy-intensive chemical transformation. How much better is impossible to say with current data, but if we use the NREL data, and subtract out these two inputs, we can get an upper estimate of 6.4 for the fossil fuel efficiency of vegetable oil. [2]

Carbon Contribution

According to the same study, replacing petrodiesel with biodiesel would reduce the total lifecycle carbon dioxide contribution by 78 percent. This reduction does not happen at the tailpipe where both fuels release about the same amount of carbon dioxide, but is because most of the carbon contained in the biodiesel and released in the process of getting the biodiesel from the crop to the fuel tank is carbon that was pulled out the atmosphere by the original oil crop, as opposed to petrodiesel where all the carbon associated with this fuel was pumped out of the ground. The total carbon dioxide produced by production, transportation, and burning of biodiesel is 679.78 grams per brake horsepower-per-hour (bhp-h). Of that, 543.34 grams is carbon dioxide that was absorbed by plants to produce the oil. That leaves 136.45 grams from fossil fuel sources.

Again, vegetable oil should reduce carbon even more because it requires less energetically demanding processing and also because it does not require methanol as an ingredient, as biodiesel does; the vast majority of methanol is produced from natural gas by refiners. The study says that the conversion of oil to biodiesel makes up about a third of the total lifecycle carbon dioxide contribution from fossil fuel sources. By leaving out this step, vegetable oil can reduce the carbon dioxide contribution by perhaps 85 percent.

Waste Oil

The above calculations are based upon the assumption that the oil is being produced for use as fuel. If the oil is a waste product, the energy costs and carbon produced up to the point where you pick up the oil from a restaurant can be treated as sunk costs. The amount of energy and carbon dioxide released in the production of the oil does not depend upon how the restaurant chooses to dispose of it. With waste oil, the only carbon dioxide you have to be concerned with is from fossil fuel energy inputs that are required to pick up the oil, process it into a usable fuel, and get it into your vehicle, and the petrodiesel you burn in order to bring the vehicle up to temperature in a two-tank system.

Tailpipe Emissions

Unfortunately, nothing definitive can be said about the tailpipe emissions of diesel engines fueled with adequately heated vegetable oil. Of the studies that have been done on the emissions of an engine burning vegetable oil, most of these did not adequately heat the oil, and the results are not valid. Of the few studies that have been done, the results vary widely, probably as a result of differences in testing protocols. From a theoretical standpoint, a diesel engine burning adequately heated vegetable oil should have emissions very similar to biodiesel. Carbon monoxide and particulates should be lower than that of diesel, and NO_X may be slightly higher.

At the end of the day, it must be recognized that no matter whether we are burning petrodiesel or vegetable oil, we are burning hydrocarbons that are very similar at a molecular level, and so the tailpipe pollution is also going to be similar. Using vegetable oil as fuel will not markedly improve the local air quality.

Destruction of Wilderness

There is a hidden cost to using any biofuel, and that is that it increases the commodity prices of oilseed crops, prompting the conversion of environmentally important "waste" lands to the cultivation of oil-producing crops. In the past few years, encouraged by high oil prices, thousands of square miles of rainforest and bogs have already been destroyed in order to cultivate oil crops in Indonesia, Malaysia, India, Brazil, Columbia, and other countries, at the cost of millions of tons of carbon Dioxide being released into the atmosphere, the loss of biodiversity, and the destruction and displacement of local communities. The expansion of oil crop plantations into wilderness is projected to increase dramatically over the next twenty years, prompted by increased demand for biodiesel.

If this concerns you, there are no easy solutions. Buying fresh oil, no matter the country of origin, increases total oil demand and raises commodity prices worldwide. Using waste oil can be much less problematic, but in the US, waste oil is sold by rendering companies as yellow grease, and to a certain extent yellow grease is a substitute for fresh oil. By increasing demand for waste oil, that raises the price of yellow grease, which in turn also raises the cost of fresh oil. In most parts of this country, the supply of used cooking oil by restaurants is only very loosely connected to the commodity price of yellow grease and a handful of people using waste oil will have no real effect. This is changing though, and in some cities, the supply of used cooking oil is tightly connected to biofuels economy.

Personally, I feel that in communities where restaurants have to pay to have their oil taken away, using waste oil as fuel is still a relatively uncomplicated environmentally responsible act, since the local oil supply has not been effectively connected to the global oil economy.

Economic Reasons

Using waste oil as fuel can save you a significant amount of money, depending upon how much you value your labor, your oil collection and processing setup, the fuel economy of your vehicle, and your driving habits.

The savings can be calculated using this formula:

$$\textit{Savings} = \frac{\textit{Miles Driven} \cdot \textit{Percent of Miles Running Vegetable Oil} \cdot (\textit{Cost of Petrodiesel} - \textit{Cost of Vegetable Oil})}{\textit{Fuel Economy of Vehicle}} - \textit{Initial Cost of Conversion and Collection and Processing Equipment}$$

As you can see, savings are dependent upon a number of factors that don't have much to do with the fuel, but instead depend upon driving habits and fuel economy. If you have a very fuel efficient vehicle that you drive for less than 5,000 miles per year, mainly for short trips during which the engine is warm enough to use vegetable oil for only a small portion of the time, then you may never save any money. If you are more like the average driver, and drive around 12,000 miles a year on trips of varying duration, and spend most of the time burning vegetable oil, you'll probably make back your initial investment of $2,000 to $3,000 after two years or so. If you are a high-mileage driver, then you can start saving money within months of conversion.

As important as fuel economy and driving habits is the real cost of vegetable oil fuel. It's easy to figure out the price of fresh oil, but figuring out the cost of waste vegetable oil can be more difficult. Obviously, included in the costs should be the electricity, disposable filter, and supplies that are part of your collection and processing setup. Perhaps less obvious, you should also probably count the labor cost that goes into collecting and processing every gallon of waste oil. Most people spend about an hour or so for every forty gallons of fuel they collect. If your time is worth $12 an hour that means the fuel costs $0.30 in labor per gallon. If your time is worth $80 an hour, the labor cost of the fuel is $2.00 per gallon.

You owe road taxes on whatever fuel you burn on public roadways, although many people don't pay these taxes on the vegetable oil they burn. If you plan on complying with the law, you should calculate the road tax into cost of vegetable oil fuel.

Warranties

The dealer will probably use the fact that you have been running vegetable oil as grounds for not honoring any warranty you have on your vehicle, even if the problem is completely unrelated to the alternative fuel. This is not legal in this country, as the Magnuson-Moss Warranty Act requires that the dealer show that the damage was actually caused by the aftermarket alterations in order to deny warranty coverage.

Even the law is on your side, expect to have to fight hard to get the dealer to do what's right. Whether or not that is worth your time, you will have to decide.

Legalities

According to the Environmental Protection Agency, it is currently illegal to convert a diesel to run on straight vegetable oil, without an exemption from that agency, because it constitutes illegal tampering of emissions-related components. To my knowledge, the EPA has never gone after individuals for illegal tampering, but if you are considering using vegetable oil in your business for a number of vehicles, I would suggest making contact with your local EPA office to ask for guidance.

Other Factors

Besides the time that it takes to process and collect the oil, you should also recognize that if you are using waste oil, this is not the cleanest activity imaginable. Even with the best possible setup, there will be spills, there will be bad smells, and you will get oil on you and your clothes on at least one occasion.

In addition to having a diesel vehicle, the money to convert the vehicle, and the time and inclination to deal with waste vegetable oil, you also need a place to store and process the oil. If you own your own home or shop, then that's not a problem. If you rent, finding a place to deal with the oil can be challenge.

Comparison of Other Alternative Energy Paybacks

Most of our customers report that within two years they saved enough money burning vegetable oil to pay for the initial investment. A two-year payback compares very favorably to other alternative energy options available on the marketplace. The typical payback for a domestic solar hot water system is five to eight years, the typical payback for photovoltaic cells is seven to fifteen years, and a Prius may never pay for itself.

Geopolitics

Many people see using biofuels as a political act, because our country's dependence on oil is seen as making us dependent upon the will of foreign nations, because our dependence on oil leads us into foreign wars at the cost of blood and treasure, or because multinational petroleum industries are seen as malevolent forces.

Certainly, at the present scale, the effect of using waste vegetable oil does have the effect of lessening, very slightly, our dependence upon foreign or corporate sources of energy. However, as biofuels such as biodiesel and ethanol continue to grow, the result will probably be that we will become dependent upon the wills of a different set of nations and somewhat different set of corporations.

Because of questions of scale, using waste vegetable oil will probably continue indefinitely to actually reduce our dependence on foreign nations or corporate interests.

Tinkering

For a certain kind of person, using vegetable oil as fuel can be an extremely pleasurable pastime. The field is very young and it is very easy for an individual to do something new, to make a significant improvement in technology, or to shift our understanding of how to use this fuel. Some people very much enjoy doing things to a vehicle that the manufacturer never intended, in making their own fuel, and in being part of a community of experimenters.

Straight Vegetable Oil vs Homebrew Biodiesel

Using straight vegetable oil shares many of the same advantages and disadvantages of making biodiesel for yourself. Both have the potential to save you money. They have similar environmental and political impacts. Using either fuel is illegal and will give you the same warranty troubles. Both require a space to muck around with and a willingness to get dirty from time to time.

A big difference is that straight vegetable oil has a high initial cost and low operating costs, and homebrew biodiesel has a low initial cost and higher operating costs. Ultimately, which one is right for you probably most depends on what kind of person you are.

Making biodiesel involves a capacity and willingness to safely handle dangerous chemicals: flammable methanol and highly corrosive catalysts. You must regularly make very careful and precise measurements and use good lab discipline, and you must test your fuel much more rigorously. If you are the kind of person that might be interested in brewing your own beer, than this might be a good fit for you.

If working on cars is more appealing to you, then straight vegetable oil may be a better fit, since the action is mainly in the vehicle, and requires much less attention to the chemistry of the fuel.

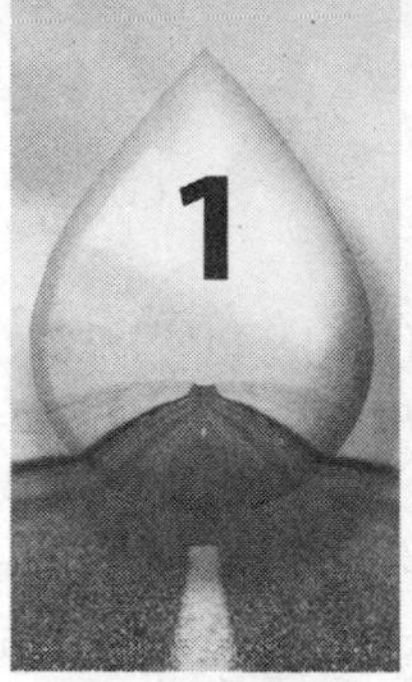

Diesel Engines

Straight vegetable oil is an excellent short-term substitute for petroleum diesel fuel, which we'll hereafter refer to as "petrodiesel." Compared to the stock fuel, conventional engines fueled with vegetable oil produced nearly the same amount of power, without consuming much more fuel — very different from nearly all other alternative fuels such as ethanol, propane, or liquid natural gas.

Unfortunately, research and experience have shown that using unheated or inadequately heated vegetable oil for extended periods causes performance to degrade dramatically, sometimes in as little 40 hours.[1] In every long-term engine study, the researchers identify the same cause for failure: carbon deposits forming on the fuel injectors, ring landings, exhaust valves, and exhaust valve stems. Some studies also show degradation of the lubricating oil caused by contamination by vegetable oil fuel.

There are five approaches that have had varying amounts of success in overcoming these problems. The approaches can be divided into two broad strategies: modifying the fuel to make it more like petrodiesel, or designing or modifying the engines to optimize them for the chemical and physical properties of vegetable oil.

Modifying the Fuel

- *Biodiesel.* Vegetable oil is chemically altered into molecules that more resemble petrodiesel. Biodiesel will burn well in nearly any diesel engine.

However, is not chemically compatible with many older fuel systems and, in these systems, can lead to swelling or failure of fuel lines, seals, and gaskets.

- *Blending.* Some conventional diesel engines can handle blends of up to 30 percent straight vegetable oil without problem.
- *Heating.* As vegetable oil is warmed, the physical properties of the fuel converge with those of petrodiesel. If the oil is adequately heated, it can be used long term as a fuel for any diesel engine. Heated vegetable oil is chemically compatible with older fuel systems.

Modifying the Engine

- *Designing an entire engine for vegetable oil.* There have been a number of engines that have been designed and built that can handle unheated vegetable oil, most notably the Elsbett engine.
- *Modifying a conventional engine.* By replacing components and adjusting injection pressure and timing, it is possible improve a conventional engine's ability to burn vegetable oil

That both strategies work underlines the fact that the success or failure of using vegetable oil as a fuel for a diesel engine depends upon understanding the *interaction* between engine design and fuel properties. We will spend the first half of the book on this topic.

First, we will discuss the aspects of a diesel engine that are critical in understanding how fuel is burned and how deposits form, and then we'll discuss in depth how the fuel properties of vegetable oil interact with those components and how those properties can be altered by heating to avoid the problems of carbon deposits.

Of the approaches above, we will focus on heating, but will discuss blending and engine modifications as well. We will not discuss biodiesel in depth, because there are already excellent resources available which address making and using biodiesel fuel.

Diesel Engines

In this section we are going to discuss those aspects of a diesel engine that determine whether it can successfully burn a fuel without forming deposits or contaminating the lubricating oil. Specifically we are going to discuss how fuel is injected, how it mixes with air, how it ignites, how it combusts, and how the combustion chamber is sealed from the lubricating oil. The engine systems that we will be discussing are the injection system, combustion chamber design, and the piston rings.

In the next section, we will discuss the fuel properties of vegetable oil extensively, but will have cause to refer to some of those properties in this section.

Basic Engine Theory

Imagine a syringe--simply a plunger and a cylinder. By moving the plunger up and down, the volume above the plunger changes. The cylinder is sealed from the outside air except by the small opening of the syringe tip. On the plunger side, the plunger fits so tightly against the cylinder walls that air is sealed from escaping past the plunger. If you draw down the plunger, the volume above the plunger increases and air or fluid is drawn into the syringe; as you push down the plunger, the volume decreases and the air or fluid is pushed out of the syringe (Fig. 1.1).

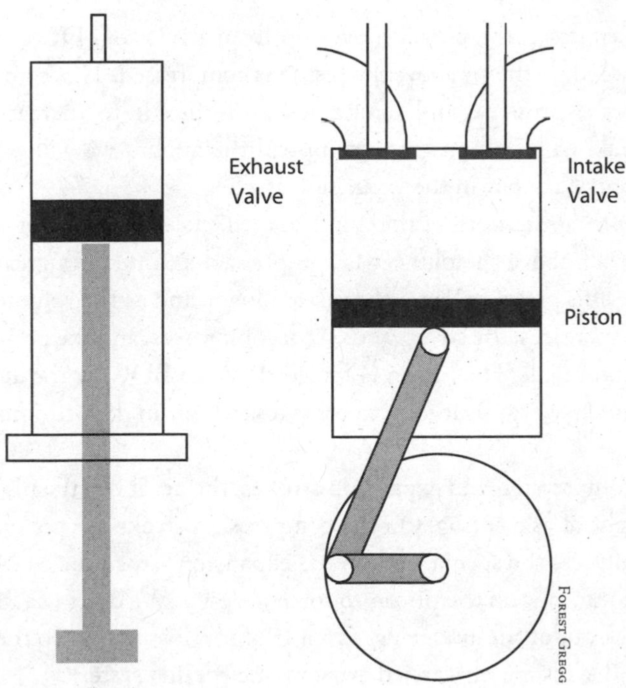

Fig. 1.1: *The plunger is the piston within the cylinder. Timed valves control the intake of fresh air and the exhaust of combustion products.*

An engine cylinder is very similar to a syringe, but with some minor differences. In an engine the plunger is called a piston, and instead of one opening, there are two valves that can open and close. One is called the intake valve and the other the exhaust valve.

Intake and Exhaust Stroke

On the *intake stroke*, the intake valve is open and the piston starts at the top of the cylinder (this is called top dead center or TDC) and moves all the way to the bottom of the cylinder (called bottom dead center or BDC), drawing in a fresh air.

On the *exhaust stroke*, the exhaust valve is open, and the combustion products are forced out of the cylinder by the piston as it moves from BDC to TDC.

When the engine is taking in air or exhausting the combustion gases, it is acting like a large air pump. Like any pump, it takes energy to move the air in and out of the cylinder.

Compression and Expansion

On the *compression stroke*, the piston moves up from BDC to TDC but both valves are closed, so the air above the piston is compressed. This compression causes the pressure and temperature of the air to increase dramatically. It takes a lot of energy to compress the air, but we will get most of that energy back out in the expansion stroke.

On the *expansion stroke*, both of the valves are still closed, the piston is at TDC and the air above the plunger is compressed and under a great deal of pressure. This pressure forces the piston down, and as the volume above the piston increases, the air expands or decompresses, and the pressure and temperature fall. The piston is forced down to BDC by the air with almost as much energy as it took to compress the air in the previous stroke.

During the compression and expansion strokes, the air above the piston can be thought of as a spring. On the compression stroke the piston does work to compress the spring, and on the expansion stroke the compressed spring does work on the piston to force it down. We don't get all of the energy back out of the air-spring, as some of it is lost as heat to the cylinder walls and is ultimately carried away by the cooling system.

All Together

The order of the strokes is intake, compression, expansion, exhaust. So,

- Intake — The piston moves from the top of the cylinder (TDC) to the bottom of the cylinder (BDC) with the intake valve open. Fresh air is drawn in.
- Compression — The intake valve closes, and the piston moves from the bottom of the cylinder BDC. The piston does work on air by compressing it. The pressure and the temperature of the air increases.
- Expansion — Both valves are still closed, the compressed air does work on the piston, pushing it down from TDC to BDC. As the volume above the piston increases, the air expands. The pressure and temperature drop.
- Exhaust — The exhaust valve opens, and the piston moves from BDC to TDC, forcing all the air out of the cylinder. At the end of this stroke, the exhaust valve closes. The next stroke is the intake.

This is the cycle of a four-stroke engine, gasoline or diesel. Two-stroke and rotary engines also have intake, compression, expansion, and exhaust as phases in their cycles, though each phase does not have its own piston stroke.

Adding Energy

Notice that in the description above, we have not burnt any fuel or added any energy to the system. So far, we have described a system that is a net energy loser, because it takes energy to pump air into and out of the cylinder, and we lose some of the energy we use to compress the air-spring to heat.

In order to get useful energy out of the engine, a gasoline or diesel engine quickly burns fuel while the piston is near the end of the compression stroke and at the beginning of the expansion stroke. The combusting fuel increases the pressure and temperature of the compressed air, which means, during the expansion stroke, the gases push down on the piston with more energy than was required to compress the air on the compression stroke.

If the added energy pushes the piston down with more energy than it took to compress the air and pump air in and out of the cylinder, then

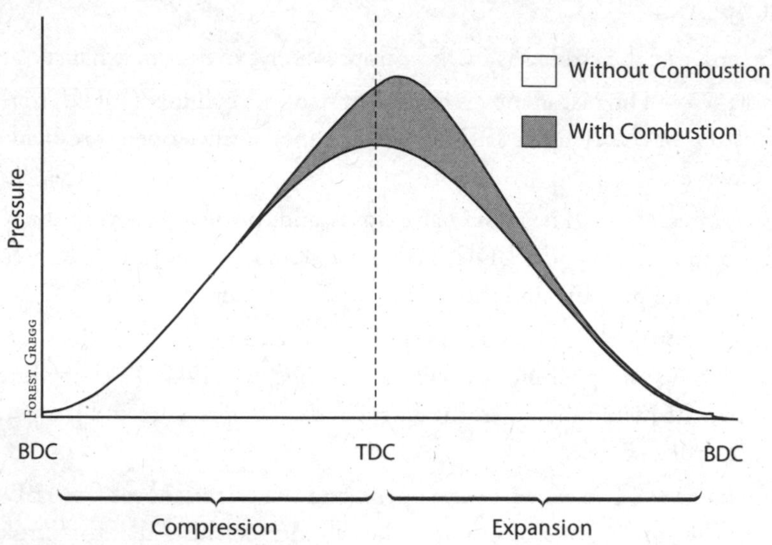

Fig. 1.2: *The addition of fuel increases the temperature and pressure of compressed air once it reaches ignition temperature.*

there will be energy left over to do useful work like moving a vehicle down the road (Fig. 1.2).

Crankshaft Position

The up and down motion of the cylinder is transformed into rotary motion by a crankshaft, which is similar in principle to the bicycle crank, which turns the up and down motion of the bicyclists legs into the rotary motion of the bicycle wheels.

The coupling of the piston to the crankshaft gives us a useful way to refer to the position of the piston. Since every stroke of the piston corresponds to 180° of crankshaft, that means that when the piston is halfway through an upstroke, we can call that position 90° before TDC (top dead center), if the piston is a third of the way down a down stroke, we can call that 60° after TDC (Fig. 1.3).

The working of an engine is inherently a dynamic process that is very difficult to capture with words or a static picture. If any of this is new information, I strongly encourage you to do an internet search for "diesel-cycle animation" and "crankshaft animation."

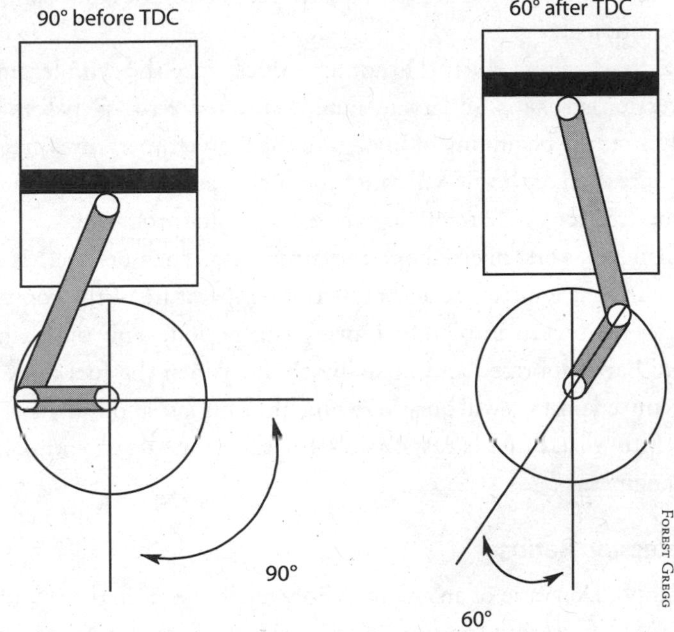

Fig. 1.3: *The image at left shows the piston moving upward in compression, at a position 90° before Top Dead Center. The image at right shows the piston moving downward after it has passed Top Dead Center, at a position 60° after TDC.*

Basic Diesel Theory

In principle, diesel and gasoline engines differ primarily in how the fuel is introduced into the cylinder, how it ignites, and by typical compression ratios. As we will see, these differences make typical diesel engines more fuel efficient than typical gasoline engines, but they also make them more expensive. In particular, the higher compression of diesel engines and high injection pressure of the fuel requires stronger and more precisely formed components, both of which have historically made diesel engines more expensive than their gasoline-powered counterparts.

Introduction and Ignition of Fuel

In a gasoline engine, fuel vapor and air are mixed together before the compression stroke. At the top of the compression stroke, the mixture is ignited by a high-voltage electrical arc jumping across the points of a spark plug. Because, the air and fuel are already pre-mixed, the combustion happens very quickly.

In a diesel engine, the fuel is not introduced into the cylinder until the compression stroke is underway. Fuel is injected 2°to 15° before TDC. Shortly after the beginning of injection, the high temperatures created by the compression of the air will cause the fuel to autoignite. Fuel continues to be injected for up 32 to 40 degrees of crankshaft rotation.[2]

When fuel is first injected into the combustion chamber there is a short delay, called the ignition delay, before the temperature of the compressed air causes the fuel to autoignite. During this period, some of the injected fuel will have vaporized and mixed with air. When the fuel does ignite, this premixed portion will burn very quickly. This fast and violent combustion of pre-mixed fuel is responsible for the characteristic knocking of diesel engines.

Compression Ratios

The compression ratio of an engine is how much the air in the cylinder will be compressed during the compression stroke. Specifically, it is ratio of the volume above the piston when it is at BDC compared to the volume above the piston when it is at TDC. The higher the compression ratio, the higher the temperature and pressure of the air will be at the top of the compression stroke (Fig. 1.4).

Gasoline engines are typically limited to a compression ratio of about 9:1, as higher compression ratios would cause the fuel to autoignite before the spark plug fired. In a gasoline engine, when this happens it's called predetonation, preignition, or knock.

Because diesel engines depend upon compression to heat the air to the point where the fuel will autoignite, the compression ratio can be, and indeed must be, higher. Indirect injection diesel engines typically have compression ratios from 18:1 to 24:1,[3] and direct injection engines have a ratio of 12:1 to 24:1.[4]

Basic Combustion Theory

Combustion is a rapid, heat-releasing reaction between fuel and oxygen — in other words, a fire. In this case, whether we are talking about diesel or vegetable oil, the fuel consists of large molecules made up of hydrogen and

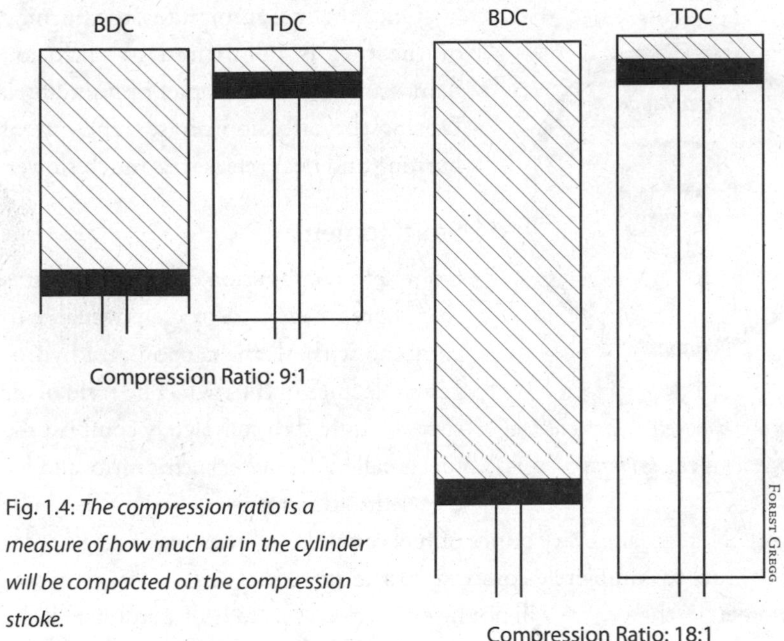

Fig. 1.4: *The compression ratio is a measure of how much air in the cylinder will be compacted on the compression stroke.*

carbon atoms, that when reacted with oxygen will produce carbon dioxide (CO_2) and water (H_2O), if the combustion is perfect and complete.

Premixed and Diffusion Burning

In a diesel engine, there are two types of combustion: premixed burning and diffusion burning. With premixed burning, the fuel has already turned into a gas and has mixed with the air, when this mixture ignites, it combusts very quickly. An example of premixed burning would be releasing the gas from a cigarette lighter into a bottle and then striking a flint. Since the fuel is already mixed with gas, it will burn very quickly.

The flames from a lighter or propane torch are examples of diffusion burning. The reaction between oxygen and the fuel is happening at the edge of the visible flame (Fig. 1.5). It can't happen inside this edge, because any oxygen that would diffuse inside flame is consumed at the area of reaction — the edge of the flame. Diffusion burning is always much slower than premixed burning and is controlled by the rate of diffusion of the fuel into the air and the air into the fuel. With liquid fuels, the rate of evaporation of the fuel is critical to determining the rate of burning.

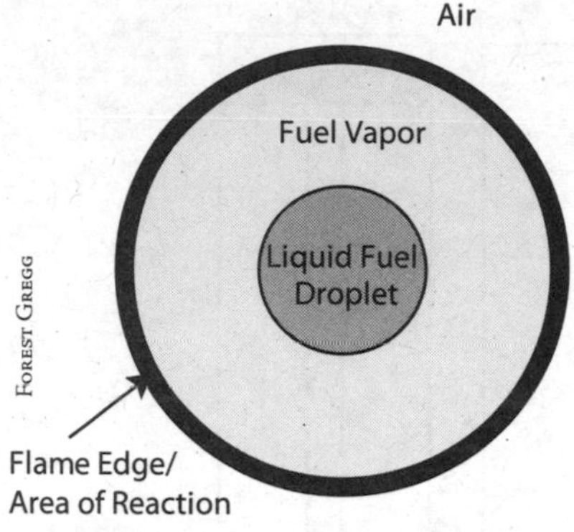

Fig. 1.5: *The reaction between oxygen and fuel occurs at the edge of the visible flame.*

Once the fuel autoignites, the burning of the fuel is dominated by diffusion flames around each droplet of liquid fuel. During the diffusion phases, the rate of burning and heat release are much slower.

Stoichiometry

Complete combustion of a fuel requires that there be enough oxygen available to combine with all the carbon and hydrogen molecules in the fuel. The ratio of air to fuel required to completely combust the fuel is called the *stoichiometric* ratio, and for diesel fuel this ratio, by weight, is about 14.4:1. That means every ounce of fuel requires 14.4 ounces of air in order for the fuel to completely combust. If the air to fuel ratio is less than stoichiometric, then there will not be enough oxygen to fully combust all the fuel, leaving some fuel unburnt. A mixture of air to fuel that is less than stoichiometric is called *rich*. A mixture of air to fuel that is greater than stoichiometric is called *lean*.

The overall air to fuel ratio in a diesel engine is always lean, but locally the mixture will vary from very lean, in areas far away from the fuel spray, to stoichiometric at the edge of flame, to infinitely rich inside a liquid droplet of fuel.

Diesel Combustion

At this point, we should now have enough background that we can focus on the quality of combustion of fuel in a diesel engine, determined by four factors: how well the fuel mixes with air, the temperature of intake air and the temperature of the cylinder walls and piston, chemical properties of the fuel, and timing

Fuel Mixing

Fuel mixing is primarily controlled by the injection system and, in some diesels, turbulent or ordered air motion caused by combustion chamber design. First, we'll discuss injection systems.

There have been two main classes of injection systems: direct injection and indirect injection.

Direct Injection

In direct injection engines, the fuel is sprayed directly into a combustion chamber in the piston. With this setup, the spray pattern of the fuel from the injector will determine to a great extent how well the fuel mixes with the air (Fig. 1.6).

The ideal spray pattern in one in which the fuel is 1) finely atomized, i.e. broken into the very small droplets, and in which 2) the spray has enough momentum to penetrate deeply into the combustion chamber.

Atomization is important because it increases the surface area of fuel that is exposed to air, which is critical since combustion in a diesel engine happens mainly at the surface of fuel droplets.

Penetration is important because in order for the fuel to burn quickly and cleanly, it needs to be exposed to as much of the air charge as possible. If the atomization was good, but the penetration was shallow, all the fuel would be concentrated in a small area, where there may not be enough air to fully burn all the fuel.

Increasing the pressure that the fuel is injected into the cylinder increases both the atomization and the penetration, and direct injection engines use very high injection pressures, between 18,000 and 30,000 psi.[5] Unfortunately, past a certain point increasing pressure reduces the

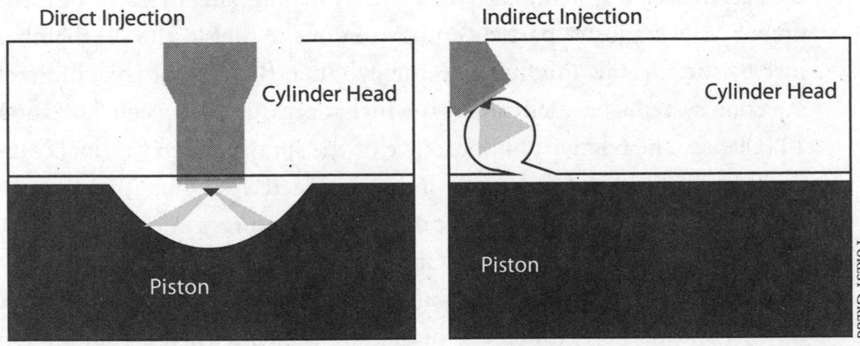

Fig. 1.6: *Direct Injection systems use very high injection pressures and rely on good spray pattern quality to be effective. Indirect Injection systems combust fuel in a prechamber, and use higher compression ratios to offset their lower efficiency.*

penetration rate, because the droplets become so finely atomized that they don't have the momentum to keep moving through the compressed air.

Spray pattern quality is also heavily influenced by the viscosity of the fuel. Thicker fuel produces sprays with a larger average droplet size, which means that the fuel will not mix with the air as well, and that diffusion burning will be slower, since the total surface area of fuel exposed to air will be less.

Because direct injection engines depend so much on the ability of the spray pattern to effectively mix the fuel and air, direct injection engines are more sensitive to differences in fuel quality than indirect injection engines.

Indirect Injection

A less common, but more descriptive name for indirect injection systems is "divided combustion chamber" systems. In these systems the fuel is sprayed into a small chamber that is thermally insulated from the cooling system and from the main cylinder, except for a small passage. The insulation means that the walls of this prechamber quickly become and remain very hot once the engine is running.

When fuel is sprayed into the prechamber, the high temperatures of the chamber cause the fuel to ignite very quickly. This ignition, in turn, forces the fuel and air in the prechamber into the main cylinder with a great deal of force that results in good mixing of fuel and air.

The spray pattern from the injectors is relatively unimportant in an indirect injection system, since the main work of mixing fuel and air is the done by the explosive partial combustion of the fuel in the prechamber forcing the fuel/air mixture into the cylinder. Because of this, indirect injection systems normally inject the fuel at pressures between 5,000 and 14,000 psi. The relative unimportance of the spray pattern for final combustion quality tends to make indirect injected engines much more accepting of a wide range in fuel properties than direct injection engines.

The high velocity of the gases in the prechamber promotes the transfer of heat from the gases to the walls of the chamber , because the higher velocity means more collisions of gas molecules with the combustion chamber walls. This heat loss lowers the pressure of the gases, reducing the amount of energy available to do work on the piston, and reducing the efficiency of indirect injection systems compared to direct injection, which generally have much calmer gas flows. This heat loss also makes it more difficult for the gases to reach the autoignition temperature when the engine is cold starting. Indirect injected engines tend to have higher compression ratios than direct injected engines to overcome this heat loss through hotter compression temperatures and to ensure good starting.

Air Motion

Some engines aid mixing through turbulent or ordered air motion in the cylinder. The basic idea is to increase the surface area of the flame front, and to prevent local areas of richness (Fig. 1.7).

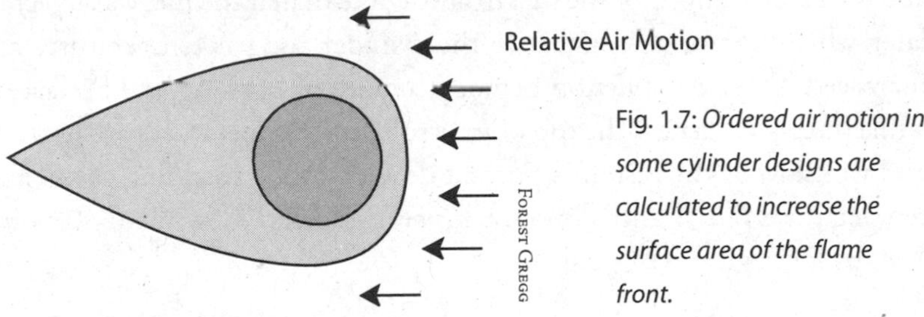

Fig. 1.7: *Ordered air motion in some cylinder designs are calculated to increase the surface area of the flame front.*

Temperature

Temperature controls the ignition delay, the rate of diffusion burning, and whether the fuel will even burn.

In order for diesel fuel to autoignite, it has to vaporize, mix with air, and go through some preliminary chemical reactions. It takes time for all this to happen, and that time is called the ignition delay. The temperature of the compressed air is the most important factor controlling how long that delay is. The hotter the air, the shorter the ignition delay.

The speed of diffusion burning is controlled by the rate of diffusion of fuel and the rate of diffusion of air. The rate of diffusion of air into the droplet is mainly controlled by the motion of air relative to the droplet, but the rate of diffusion of fuel into the air is controlled by the rate of vaporization of the fuel, which in turn is a function of the temperature. The higher the temperature, the faster the fuel will evaporate and the faster it will diffuse into the air where it can burn.

The cooler the temperature, the slower the fuel will evaporate, until the rate of vaporization is no longer high enough to maintain burning, and the flame will be extinguished. This transition point is called the flash point.

The temperature of a cylinder is determined mainly by 1) the position of the piston in the cylinder, 2) the compression ratio, and 3) the engine speed, load, and the rate of heat transfer from the walls of the cylinder and piston.

Piston Position

As we discussed previously, the compression of the air in the cylinder by the piston during the compression stroke heats the gases in the cylinder, and the expansion of the gases pushing down the piston during the expansion stroke cools the gases. On the compression stroke, if fuel is injected while the piston is still far from TDC, the gas temperatures are cooler, and it will take longer for the fuel to autoignite than if the fuel was injected later, when the piston is higher in the cylinder and gas temperatures are increased. Once combustion begins, the rate of burning will be fastest while the piston is near the top of its stroke and temperatures are hottest, and will slow as the piston is pushed down by the expanding gases, and temperatures fall. Eventually, the temperature falls below the flash point and the flames are extinguished.

Compression

Compared to a lower-compression engine, the piston in a higher-compression engine compresses the air into a smaller relative volume, not just at the top of the stroke, but throughout the compression stroke. This means that in a higher-compression engine, the piston position during the compression stroke, as indicated on the crankshaft, will be associated with higher gas temperatures than a piston at the same indicated crankshaft position in a lower compression engine.

The opposite is true on the expansion stroke. In a higher-compression engine, as the piston moves down, the relative volume above the piston increase more quickly than in a lower compression engine, so, in a higher-compression engine, gas temperatures drop more quickly in relation to crankshsaft position on the expansion stroke.

Surface Temperatures and Load

The gas temperatures are also affected by the temperatures of the piston and cylinder walls. The temperatures of these components are a function of how much heat is being imparted to them and how much of that heat these components can transfer to the engine's cooling system. The gas temperatures will be hottest during high-load, high-rpm conditions, and coolest during low-load, low-rpm conditions.

The amount of heat imparted to these components is controlled by the engine load. When the engine is under full load, the maximum amount of fuel is being injected into the cylinder, and the peak temperatures are at their highest.

Engine speed and component design control the rate that these components transfer heat to the cooling system. When the engine is turning over quickly, there is less time for the components to shed the heat. The use of different materials also controls how fast heat is shed to the cooling system, and modern diesels are incorporating materials that allow for less heat transfer.

Combustion in diesel engines is usually more efficient under high load and high speed conditions because of relatively less energy-stealing heat transfer. However, mechanical losses of high-speed operations usually negate these benefits.

Fuel Properties

We've already discussed some of chemical and physical properties of diesel fuels that affect combustion: autoignition properties, rate of evaporation, and flash point. We will be discussing these and other properties in more detail soon.

Timing

Bringing all the other elements together is timing — controlling when and how much fuel will be injected in relation to the position of the piston. The aim of timing is to reduce the amount of fuel that is burned during the premixed phase (to reduce diesel knock), have peak pressures develop as close to TDC as possible, and limit the amount of unburned fuel and other pollutants in the cycle.

The conventional terms for discussing adjusting timing are *advance* and *retard*. Advancing means having an event happen earlier. Retarding means having an event happen later. For mechanically controlled injection systems, the most critical variables that can also be easily adjusted are the beginning and end of injection.

Advancing the beginning of injection results in a longer ignition delay, because the fuel begins to be injected while the piston is lower in the compression stroke and gas temperatures are cooler. The longer delay means that more fuel has a chance to evaporate and mix with air before ignition occurs, which means that when the fuel does ignite the combustion of the premixed fuel will produce higher peak pressure and temperatures, as well as a louder knock. The higher peak temperatures lengthen the period of time during which fuel has an opportunity to burn during the expansion stroke, and therefore reduces the amount of unburnt fuel exhausting from the cylinder as smoke, or forming of deposits in the cylinder. Conversely, retarding the beginning of injection has the opposite effect.

The timing of the end of injection is critical for controlling smoke and the buildup of carbon deposits. If fuel is injected too late during the expansion stroke, the fuel will not have time to vaporize and combust before the gas temperatures fall below the point where combustion can be sustained. The unburned fuel either leaves the cylinder through the exhaust valve as smoke, or forms carbon deposits in the cylinder, typically on the exhaust valve, exhaust valve stem, ring landings, or injector.

With modern electronically controlled injectors, finer control of timing is possible. Not only can the beginning and end of injection be controlled, but the amount of fuel, as well.

Carbon Deposits

When fuel is introduced into a cylinder, but fails to completely burn, the unburned portion either is forced out of the cylinder during the exhaust stroke as smoke, or it stays in the cylinder in the form of carbon deposits. A certain level of carbon deposit is unavoidable and even acceptable, but if the deposits build up they can cause injector sticking, valve sticking, cylinder liner scuffing, and even piston seizure. Excessive deposits are a special problem when using vegetable oil, and so require critical examination.

Carbon deposits form on surfaces that are cool enough that fuel or partially burned fuel can settle on them, but hot enough to cause the fuel to boil, polymerize, or crack. If a surface is too hot, fuel or partially burned fuel cannot settle on it. Imagine a small droplet of water falling onto a very hot pan. The droplet will not wet the pan, but instead will form a bouncing sphere of water that will quickly disappear as it evaporates. What is happening with the water droplet is that as the droplet nears the surface of the pan, the surface of the droplet begins to boil, producing a cushion of steam. This cushion of steam prevents the droplet from reaching the surface, and slows the evaporation of the water droplet, since steam transfers heat much more slowly than direct contact with the surface of the pan. As the steam diffuses away, the cushion shrinks, and the droplet gets closer to the surface of the pan, which causes more water to boil and replenish the steam cushion.

The same thing happens with petrodiesel or vegetable oil fuel. If the surfaces inside the cylinder are hot enough, then they cannot be wetted by the fuel. If they are cool enough to be wetted, fuel hitting the surfaces in the cylinder will form carbon deposits if the surfaces are hot enough to cause the fuel or partially burned fuel to boil, crack, or polymerize.

Boiling

Neither petrodiesel nor vegetable oil has one boiling point, the way that water (which is just H_2O molecules) does. Instead, these fuels have a boiling temperature range. The lighter components of the fuel will begin to boil at one temperature, but the heavier components won't begin to boil until temperatures are much hotter. If the temperature of a surface is hot enough to cause the lighter components to boil, but is below the boiling temperature of the heavier components, what is left is thicker and heavier, and impurities in the fuel that won't combust or cannot evaporate will come out of solution.

Cracking

At high enough temperatures, fuel molecules will begin to spontaneously split apart into smaller, highly reactive fragments. If these fragments do not burn, they will clump back together to form very large molecules that will not evaporate. Cracking can only occur in areas where there is not enough oxygen for combustion to occur, since if the fuel is hot enough to crack, it is hot enough to burn.

Polymerizing

High temperatures can also lead to certain kinds of reactions that cause fuel molecules to link together directly to form very large molecules called polymers. Different fuels vary in how susceptible they are to polymerization and cracking reactions.

Controlling Carbon Deposits

Carbon deposits can be controlled by (1) improving the completeness of combustion and thereby limiting the amount of fuel or partially burned fuel available to form deposits, and (2) controlling the temperature of the surfaces in the cylinder: either heating the surfaces past the point that fuel can land on them, or cooling them down past the point that liquid fuel will form deposits.

Contamination of the Lubricating Oil

Ideally, the piston fits so tightly in the cylinder that it is impossible for gases to bypass the piston and enter into the crankcase. The components responsible for making that seal are the piston rings. However, the piston rings cannot form a seal when the engine is cold because that would mean that they would be too tight once the engine was at operating temperature, and the piston would seize. So this means that when a diesel engine is cold, some fuel is always going to bypass the rings, enter the crankcase, and contaminate the lubricating oil. In the next section we will discuss why unheated vegetable oil tends to impinge on the cylinder walls, which leads to more fuel bypassing the rings, and why contamination of the lubricating oil by vegetable oil is so much more dangerous than contamination by petrodiesel.

How Things Fall Apart

We have now introduced enough information about diesel engines to be able to understand how inadequately heated or modified vegetable oil causes long-term problems in conventional diesel engines. In this section, we'll address the fuel properties of unheated vegetable oil that lead to the typical syndrome of engine performance degradation and damage — carbon deposits forming on the fuel injectors, ring landings, exhaust valves, and exhaust valve stems, and in some studies, contamination of the lubricating oil by vegetable oil fuel.

Higher viscosity of unheated vegetable oil causes the fuel spray to consist of larger droplets, which reduces the efficiency of air and fuel mixing, resulting in local areas of richness. The large droplet size also decreases the overall surface area of fuel, slowing the rate diffusion burning. To review:

1. Higher boiling temperatures slow down the rate of evaporation and slows the rate of diffusion burning.
2. Higher bulk modulus, which we'll discuss in Chapter 2, causes mechanically controlled injection systems to open earlier and close later, as well as creating nozzle dribble or unintended secondary injection.
3. Higher flash point shortens the period of time that cylinder temperatures are hot enough for the fuel to burn.
4. Higher oxidative instability makes vegetable oil much more likely to form carbon deposits if unburned fuel can land on surfaces.

Putting it altogether, when unheated vegetable oil is used to fuel an unmodified conventional diesel engine, the fuel combusts less efficiently because it does not mix as readily with air, and the slower rate of combustion and higher flash point ensure that the mixture does not have sufficient time to burn. The loss of combustion efficiency is enough to cause some carbon deposits to form, but the problem is aggravated by nozzle dribble or secondary injection, which results in fuel being introduced into the combustion chamber after temperatures have already fallen past the point where fuel can be burnt efficiently. The higher oxidative instability of vegetable oil compounds the problem of all this unburned fuel turning to carbon deposits.

In Chapter 2, we will discuss those properties of vegetable oil which make it an excellent substitute for diesel fuel, if the long-term engine performance problems can be overcome.

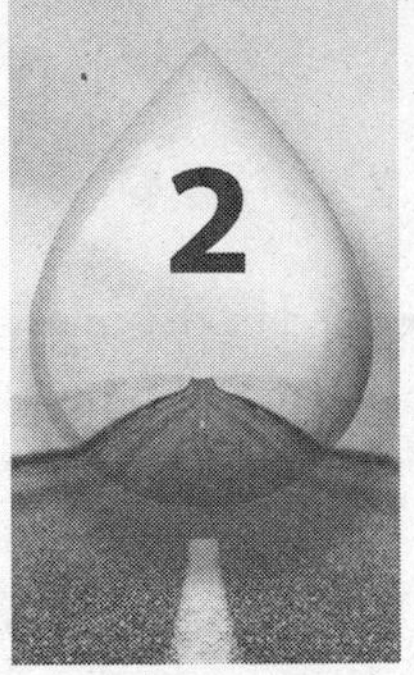

Fuel Properties

What is Vegetable Oil?

For our purposes, vegetable oil (1) consists of molecules called triglycerides, (2) is liquid at room temperature, and (3) is derived from plants. Triglycerides are molecules consisting of three long fatty acids connected to a backbone called glycerol or glycerin (Fig. 2.1). Solid fats, from animals or plants, are also triglycerides, and can be used as diesel fuel, although their nonliquid nature adds complications.

Plants mainly produce triglycerides as an efficient way to store energy. By volume, vegetable oil contains more than three times the energy of sugar or starch. The energy is stored in the numerous carbon-carbon and carbon-hydrogen bonds of the fatty acids.

Fatty Acids and Differences Between Vegetable Oils

Vegetable oil from one species of plant can be radically different from oil produced by a different species, or even strain. For example, coconut butter is semi-solid at room temperature and has an indefinite shelf life, whereas linseed oil is liquid at room temperature but will "dry" into a thick gelatinous, foul-smelling mass within a day or so of exposure to air. The difference between these two vegetable oils is the fatty acids.

Plants can make about twelve different kinds of fatty acids, which differ in the length of the carbon chain and the number of double bonds in the chain. Every species and strain of oil crop varies in how much the different fatty acids are typically expressed in the oil, and this fatty acid profile determines properties like viscosity, melting point, chemical stability, and high temperature behavior.

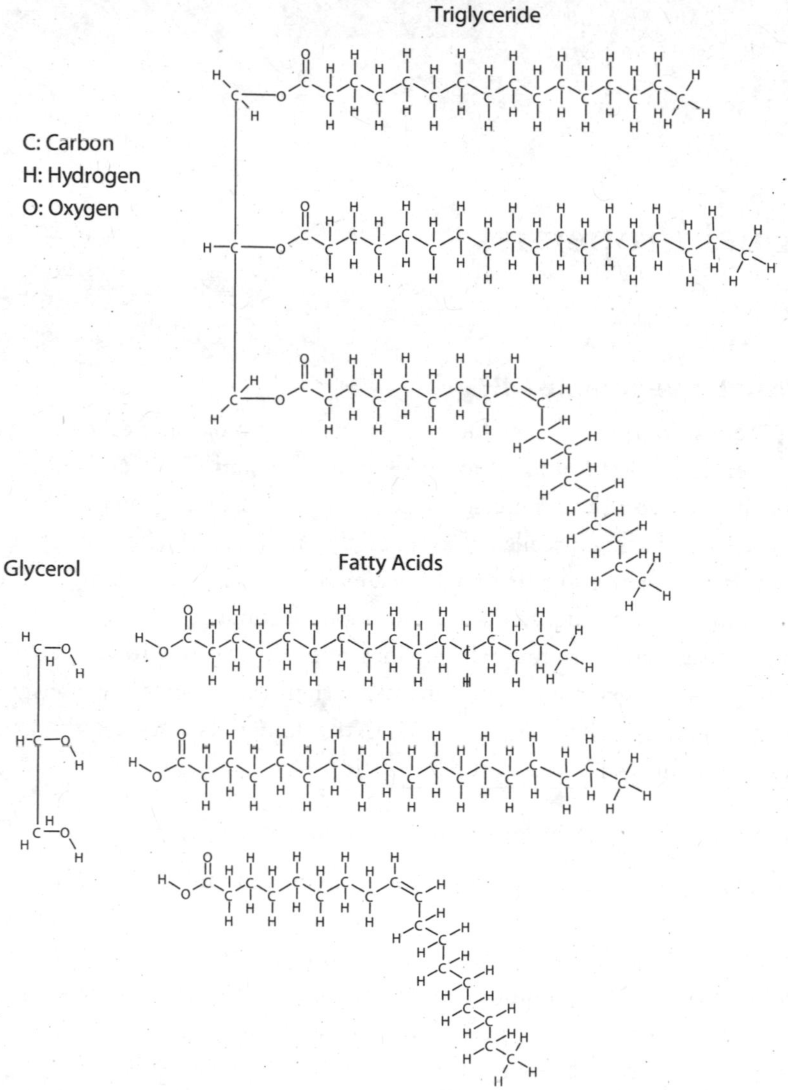

Triglyceride

C: Carbon
H: Hydrogen
O: Oxygen

Glycerol

Fatty Acids

Forest Gregg

Fig. 2.1: *Triglycerides are molecules consisting of three long fatty acids connected to a backbone called glycerol or glycerin.*

Chain Length

Vegetable oil fatty acids vary in the length of their carbon chain from 14 to 24 carbon atoms. In general, oils that contain more, and larger fatty acids will have a higher melting point, be thicker, and contain more energy by volume.

Saponification value is an indicator of average chain length. The higher the saponification value the smaller the average chain length.

Double Bonds

Fatty acids can also vary in their number of double bonds that exist between carbon atoms. Vegetable oil fatty acids can have 0, 1, 2, or 3 of these types of bonds.[1] The number of carbon-carbon double bonds is also called degree of unsaturation because the carbon atoms in a double bond each have one less bond with a hydrogen atom than if there was just a single bond between them. In other words, they are not saturated with hydrogen bonds (Fig. 2.2).

Naturally occurring double bonds in vegetable oil put a kink into the fatty acid chains. This kink is illustrated in Figure 2.1: the top two fatty acids do not have any double bonds and are pretty straight; the bottom fatty acid has a double bond and a bent shape.

A high degree of unsaturation explains why vegetable oil is liquid at room temperature and lard or butter is solid. The triglycerides in lard and butter contain mainly straight, saturated fatty acids, which allow them to pack together more compactly and form stronger intermolecular bonds

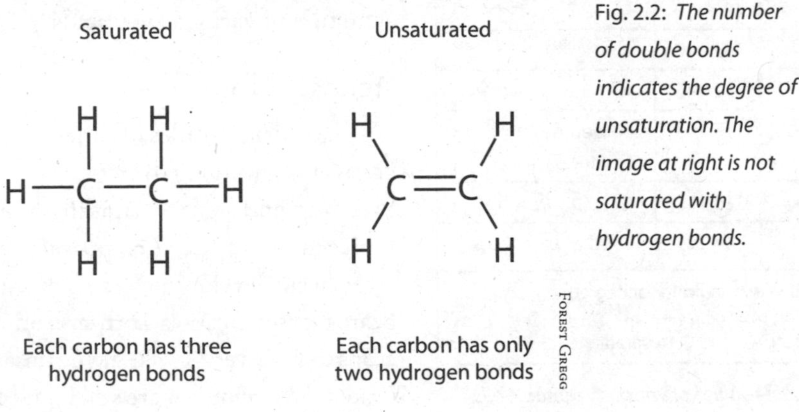

Fig. 2.2: *The number of double bonds indicates the degree of unsaturation. The image at right is not saturated with hydrogen bonds.*

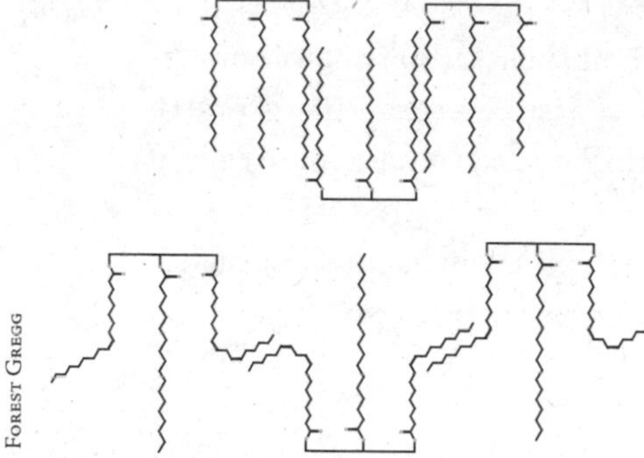

Fig. 2.3: *The triglycerides in lard and butter contain mainly straight, saturated fatty acids which can pack together compactly. Vegetable oil triglycerides contain kinked, unsaturated fatty acids which do not pack together to form strong intermolecular bonds, hence the vegetable oil is liquid at room temperature.*

Coconut	6-12	Solid at room temp
Palm	35-61	Semi solid at toom temp.
Olive	75-94	
Castor	82-88	
Peanut	80-106	
Rapeseed	94-120	
Cottonseed	90-140	
Sesame	104-120	
Corn	103-140	
Sunflower	110-143	
Soybean	117-143	Sometimes used as a paint base
Safflower	126-152	
Linseed	168-204	Used as paint base
*Iodine value is an indicator of degree of unsaturation of an oil. Source: Babu, A.K. and G. Devaradjane		

Table 2.1: *Iodine value* of various vegetable oils.*

than is possible for vegetable oil triglycerides that contain many kinked, unsaturated fatty acids (Fig. 2.3). Among vegetable oils, the greater the degree of unsaturation, the lower the melting point, the thinner the oil, and, as we will see, the less chemically stable the oil is. Iodine value is an indicator of unsaturation. The higher the iodine value the greater the degree of unsaturation (see Table 2.1).

Typical Fatty Acid Profiles

Appendix D shows typical fatty acid profiles for eleven oil crops. Below the names of the fatty acids, the first number gives the number of carbons in the fatty acid. The second number indicates the degree of unsaturation. As we proceed through the book, and discuss how different vegetable oils differ in fuel properties, return to this table for the ultimate explanation of the differences between one oil and another. See Appendix D for the structure of various fatty acids.

Trends in Crops

The fatty acid profiles of different oil crops are converging through selective breeding and genetic modification. Universities and seed companies are intensively developing strains of soybean, rapeseed, canola, corn, and other crops that are very low in highly unsaturated fatty acids but are still liquid at room temperature. This work has been going on for decades, but has been spurred recently by public concern with *trans* fats, which has created a large demand for oils that "naturally" have excellent frying properties and shelf life.

Vegetable Oil and Cooking Oil

The only types of vegetable oil commonly used for fuel in this country are cooking oils — vegetable oils produced and refined as a medium to impart heat to food through deep-fat frying, pan frying, or sautéing. Flavoring oils like sesame, walnut, and extra-virgin olive oil and rarer essential oils from rosemary or peppermint contain too many impurities for fuel use, even if the price of these oils were not so prohibitively expensive. Yellow and brown grease, which are the products produced by the rendering companies that own the waste oil dumpsters behind restaurants, have not proven acceptable as fuel, nor has crude, unrefined vegetable oil, although work continues to find a way to use these oil sources. In this book we will be primarily interested in fresh and used cooking oil and will generally use *vegetable oil* and *cooking oil* as interchangeable terms.

From Seed to Store

Seeds are the parts of plants that usually contain the highest oil content, compactly supplying the energy needed for germination. Most commercial vegetable oil is derived from the oil-rich seeds of soybeans, canola, rape, corn, sunflowers, peanuts, safflower, copra and other plants (see Table 2.2). Some plants bear fruit with high oil content in order to attract animals that will help disperse the seeds. Of this type, olive, palm, and coconut are commercially important examples.

	US	World
Soybean	2,483	6,048
Canola*	276	-
Rapeseed*	-	1,329
Corn	241	-
Palm	196	-
Coconut	127	-
Cottonseed	98	1,182
Lard	96	-
Palm kernel	86	292
Olive	70	-
Sunflower	62	819
Peanut	39	877
Safflower	13	-
Sesame	3	-
Copra	-	144

** Canola is a strain of Rapeseed popular in North America. Rapeseed is the dominant oil crop of Europe.*
Source:
http://usda.mannlib. cornell.edu/
http://usda.mannlib. cornell.edu/
MannUsda/viewDocumentInfo.do?
documentID=1289

Table 2.2: *US and World Consumption Of Edible Oil, in Millions of Gallons.*

The standard process for extracting oil starts with drying the seeds or fruit, cleaning it of impurities, then pulverizing and shredding the plant matter into a meal. At this point, process can vary depending on the end product.

Extra-virgin olive, sesame, walnut, and other oils that are valued for their flavor are pressed without heating, i.e. cold-pressed, producing a light and flavorful oil. In addition to being very expensive, cold-pressed oils are unsuitable for frying because the same components that give the oils their distinctive flavor tend to begin to burn and develop off-tastes at high temperatures. These impurities also generally make cold-pressed oils unsuitable as fuel.

The vast bulk of commercial vegetable oil is cooking oil — oil that is primarily used as medium to impart heat to food. In this country, refining vegetable oil is a very competitive and consolidated business[2] that has perfected capital-intensive extractive and refining processes that achieve maximum yield from the meal and produces cooking oil nearly free from impurities.

For cooking oil, if the meal underwent a first cold pressing, it will be pressed again while the meal is heated. After the heated pressing, a solvent is added, usually hexane, and the oil is pressed again and the hexane-vegetable oil mixture extracted. The hexane is subsequently then removed from the oil, usually below the limits of measurability of one part per 100 million.[3] Some oils, particularly soybean, are only extracted through one solvent-based extraction.

Hot pressing and solvent extraction dramatically increase the yield of oil, but also results in bitter, smelly, unappetizing oil which must undergo a series of chemical and physical processes before being commercially acceptable: degumming, neutralization, washing, drying, bleaching, filtration, and deodorizing processes which remove free fatty acids, natural gums called phospholipids, pigments, and volatile chemicals. Citric acid or another oxidation inhibiting agent is often added at the refining stage to prolong shelf life.

What results is one of the most chemically pure consumer products available. Refined cooking oil is over 99 percent triglycerides. The reason for all this work is that the contaminants that are removed can either impart off-flavors to the food, lower the temperature that the oil begins to smoke, give the oil a commercially disagreeable color, or shorten the shelf life of both the oil and the food cooked with the oil.

After refining, vegetable oil goes by rail car to food processors that use the oil in food manufacturing and to bottlers who bottle the oil for restaurants and consumers. The typical restaurant will buy their oil directly from a bottler in plastic jugs weighing 35 pounds, and containing about 4.6 gallons.

While the level of technical processing of cooking oils may be alarming, it makes life much simpler for those of us who want to make use of it as a fuel. We need only concern ourselves with contamination and changes in the oil caused by cooking and the natural degradation of triglycerides — a fairly complicated task, but nothing compared to what we would face if we had to deal with the endlessly complex cocktail of compounds naturally present in unrefined oil.

Waste Vegetable Oil

Throughout this book we will generally be making the assumption that used cooking oil is the same thing as cooking oil, just with many contaminants that must be removed. However, when appropriate, we will discuss differences between used and fresh cooking oil.

A Word About Hydrogenation

If you start collecting waste oil from restaurants, you will soon come across hydrogenated oil, probably in form of a product called Creamy Liquid Shortening. Hydrogenated oil is oil that has gone through a chemical process to reduce the degree of unsaturation in order to prolong the usable life of the frying oil.

Hydrogenated oil is quickly becoming unpopular for health reasons. In addition to transforming double bonds to single bonds, the hydrogenation process changes the shape of some double bonds into what is called a *trans* bond, which results in *trans* fat, which have been shown to increase the risk of heart disease.

Hydrogenated oil will burn fine as fuel, as long as it is adequately heated. Hydrogenating an oil dramatically increases the melting point of the oil, which means that solid fats can clog the fuel system, unless the entire vegetable oil fuel system is heated. At Frybrid, LLC, where we've made hundreds of conversion kits, the entire vegetable oil fuel system *is* heated and we have had no problem using hydrogenated oils as fuel.

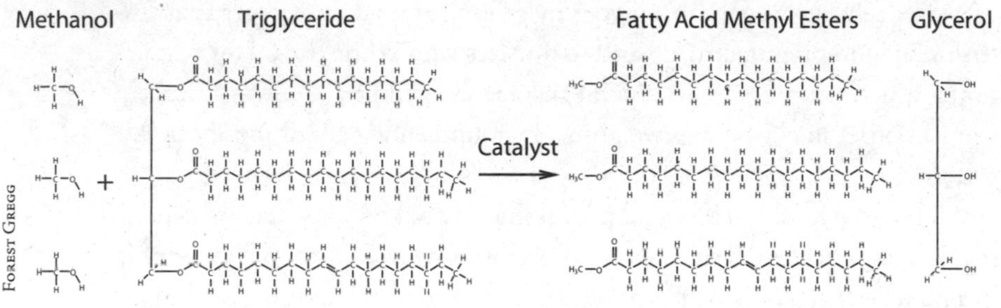

Methanol
Triglyceride
Catalyst
Fatty Acid Methyl Esters
Glycerol

Forest Gregg

Fig. 2.4: *The transesterification process produces fatty acid methyl esters and free glycerol.*

Vegetable Oil is Not Biodiesel

Biodiesel is vegetable oil that has been chemically altered into a new and much thinner product, though one that shares some characteristics with vegetable oil. Basically, biodiesel is made by disconnecting the three fatty acid chains and thereby reducing the average size of the resulting molecules to a third of the size of the original triglyceride. That reduction in size thins out the viscosity of biodiesel to a level comparable to petrodiesel.

A Bit Deeper

As we mentioned earlier, a triglyceride is made up of three fatty acids connected to a backbone molecule called glycerol. Glycerol, also called glycerine, is an alcohol[4], and the bond that connects the glycerol and the fatty acid is called an ester bond, and is the typical way that all alcohols and acids connect. The reaction that produces biodiesel is just the replacement of the alcohol portion of the ester bond with a simpler alcohol, usually methanol. This is called transesterification. In addition to the new esters that are formed called fatty acid methyl esters, transesterification also produces free glycerol, which separates out and must be removed (see Fig. 2.4).

Heat of Combustion

The potential energy a fuel can provide is measured by heat of combustion: the amount of heat released when a given amount of fuel completely combusts with oxygen. For fuels, heat of combustion is usually given as the lower heating value.

Vegetable oil has about 89-98 percent of the energy by volume of petrodiesel. The slightly lower energy content results in somewhat lower miles per gallon, slightly diminished peak power at wide open throttle, and a slower idle speed.

The lower heating value for petrodiesel is around 129,000 BTU per gallon. The lower heating value for common vegetable oils range between 115,000 and 127,000 BTU per gallon[5,6,7,8,9,10,11,12,13,14,15,16,17]. In other words, vegetable oils have between 89 to 98 percent of the energy of diesel by volume. Waste vegetable oil tends to be in the lower end of the range, probably due to the fuel already being partially oxidated.

Effects on Miles per Gallon and Power

All else being equal, and knowing the relative energy content of vegetable oil compared to petrodiesel, we can estimate reductions in miles per gallon and power. If a diesel engine gets 22 miles per gallon on petrodiesel, it should get from 89 to 98 percent of that using vegetable oil, or 21.6 to 19.6 miles per gallon.

Similarly, if top-end power is limited by the volume of fuel the injection system can deliver, there should be a 2 to 11 percent decrease in top-end power when the engine is at wide open throttle. Under normal driving conditions, a driver will not notice a loss in power because they will unconsciously compensate with the pedal for the less energetic fuel.

Finally, in systems where idle RPM is achieved by setting idle fuel delivery, a diesel engine will idle at a lower RPM on vegetable oil than diesel, because of the lower energy content of the vegetable oil fuel. This is probably a major source of the common observation that diesels are quieter running on vegetable oil than diesel.

Deep Similarity

As an alternative fuel, vegetable oil is outstanding in how similar it is to petrodiesel in the energy it holds by volume. Other alternative fuels such as ethanol, liquid natural gas, or liquid propane gas, only contain about 68 percent, 63 percent, and 72 percent respectively of the energy by volume of the fuel they replace.[18]

The similarity in energy content arises from a deep, molecular similarity between vegetable oil and petrodiesel. An average molecule of petrodiesel and a typical fatty acid consist of similar numbers of energy storing carbon-carbon and carbon-hydrogen bonds arranged in nearly the same shape (Fig. 2.5).

Fatty acids are, of course, different from petrodiesel. They are, on average, two carbons longer, which increase energy content; they contain oxygen, which decreases energy content; and, they are attached to two other fatty acids through the glycerin backbone, which doesn't have much net effect.

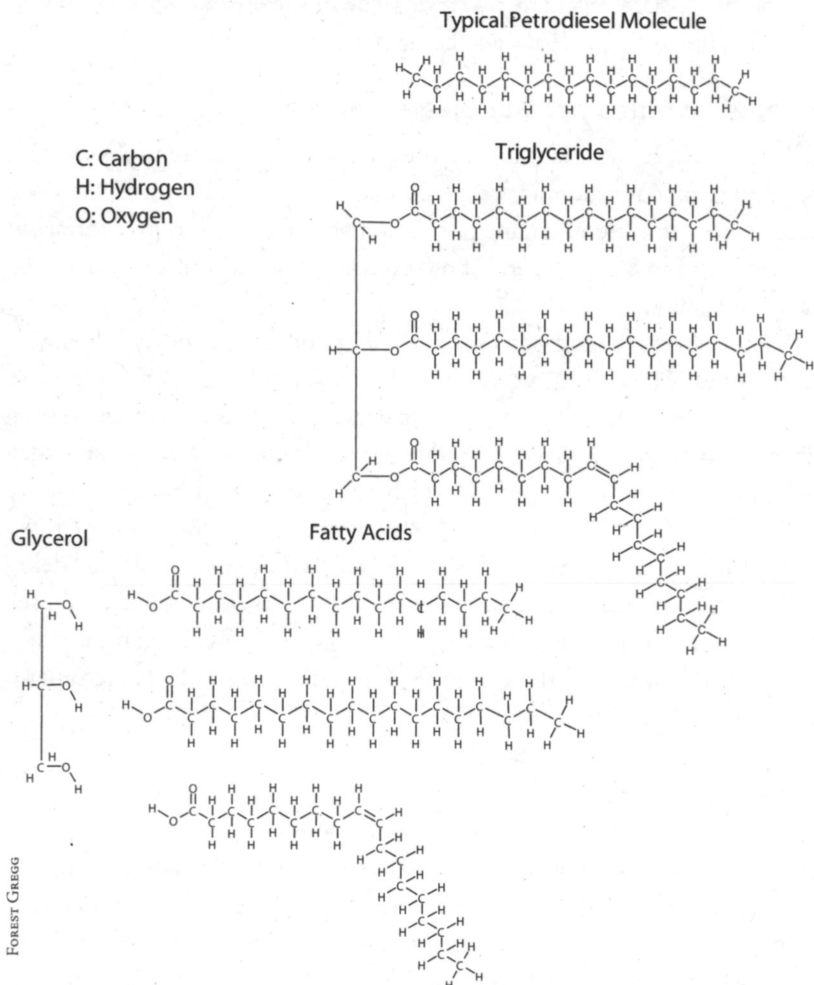

Fig. 2.5: *There is a molecular similarity between vegetable oil and petrodiesel. A typical molecule of petrodiesel and a typical fatty acid consist of similar numbers of energy-storing carbon-carbon and carbon-hydrogen bonds arranged in nearly the same shape.*

Viscosity

As a substitute fuel, the most unappealing difference between vegetable oil and petrodiesel is that common plant oils are more than ten times thicker, or more viscous, than diesel. Unthinned vegetable oil has been clearly shown to be unacceptable as anything but an emergency substitute, causing significant degradation of engine performance and quickly leading to engine damage through carbon buildup and contamination of the lubricating engine oil.

Vegetable oil can be thinned to an acceptable level by either heating the oil to at least 160°F or by chemically transforming vegetable oil into biodiesel.

What is Viscosity?

Imagine that you have three squirt guns. You fill one with water, the second with ketchup, and the third with honey. You don't have to pull the plunger very hard to get water to come out of the first squirt gun. With the second gun, you have to pull much harder to get the ketchup to squirt, and harder still to get any honey whatsoever out of the third squirt gun. How hard you have to pull to get the same volume of these various liquids is determined by the viscosity of each fluid. That's the definition of viscosity: The resistance that a fluid has to being moved. Viscosity is usually measured in either centistokes or poise, though centistokes is the standard measure for fuels. Both units can be thought of as a measure of how much force must be applied in order to move a liquid.

Viscosities of Vegetable Oils

The viscosity of diesel fuels is typically measured at 40°C or 104°F. At this temperature, petrodiesel is required to have a viscosity between 1.9 and 4.1[19] centistokes and biodiesel must have a viscosity between 1.9 and 6.0.[20] Common vegetable oils will have viscosities between 30 and 40 centistokes, over 10 times thicker than the standard for petrodiesel.

Problems of Viscosity

Pumps, Filters, and Fuel Lines. Fuel pumps have to work harder to move a more viscous fuel through a fuel system. If the oil is too thick, the pump may not be able to move enough fuel and starve the engine of fuel. Even if the pump is strong enough to move thick oil, the greater pressure or vacuum necessary to move the oil may cause problems elsewhere. Under pressure, fittings can leak and filters may rupture. Under vacuum, air leaking into the fuel is a very common problem of vegetable oil fuel systems that can lead to fuel starvation, incomplete combustion and smoky exhaust.

Atomization. If thick vegetable oil makes it to the fuel injectors, the high viscosity will produce a spray that will slowly and unevenly evaporate. The uneven spray leads to incomplete combustion and liquid oil hitting the cylinder walls, both of which will lead to coking of injectors, injector needle sticking, coking of the intake and exhaust valves, coking on the exhaust valve stem, deposits on the of the compression ring groove, and liquid fuel bypassing the rings and causing contamination of the crankcase oil.[21,22,23,24,25] See the section in Chapter 1 on fuel atomization for more discussion on the importance of spray in diesel engines.

How to Lower Viscosity

In order to successfully run a conventional diesel engine on vegetable oil, the viscosity of the fuel must be reduced. This can be done by chemically altering the fuel or simply by heating it.

Chemically altering the fuel to produce fatty acid methyl esters, also known as biodiesel, is presently the most popular strategy, as we mentioned earlier.

Heating the oil can adequately reduce the viscosity of vegetable oil to the point where it's useable as a diesel fuel. Temperature has a strong impact on the viscosity of vegetable oils, with viscosity usually decreasing exponentially with rising temperatures (Fig. 2.6).

Blending vegetable oil with another thinner fuel has been tried as a strategy to overcome the problems of viscosity, and has been generally unsuccessful. The bulk of the literature says that blends consisting of more than 20 percent vegetable oil will result in engine failure.[26,27,28,29] Whether or not blending below 20 percent will cause long term engine problems is unclear.

How Hot?

Vegetable oil only reaches 4.1 centistokes, the thickest allowed viscosity of No. 2 diesel, at around 285° F, and some important researchers suggest that this is the temperature to which vegetable oil fuel should be heated.

Fig. 2.6: *Temperature Dependence of Viscosity of Various Vegetable Oils and Diesel Fuel*

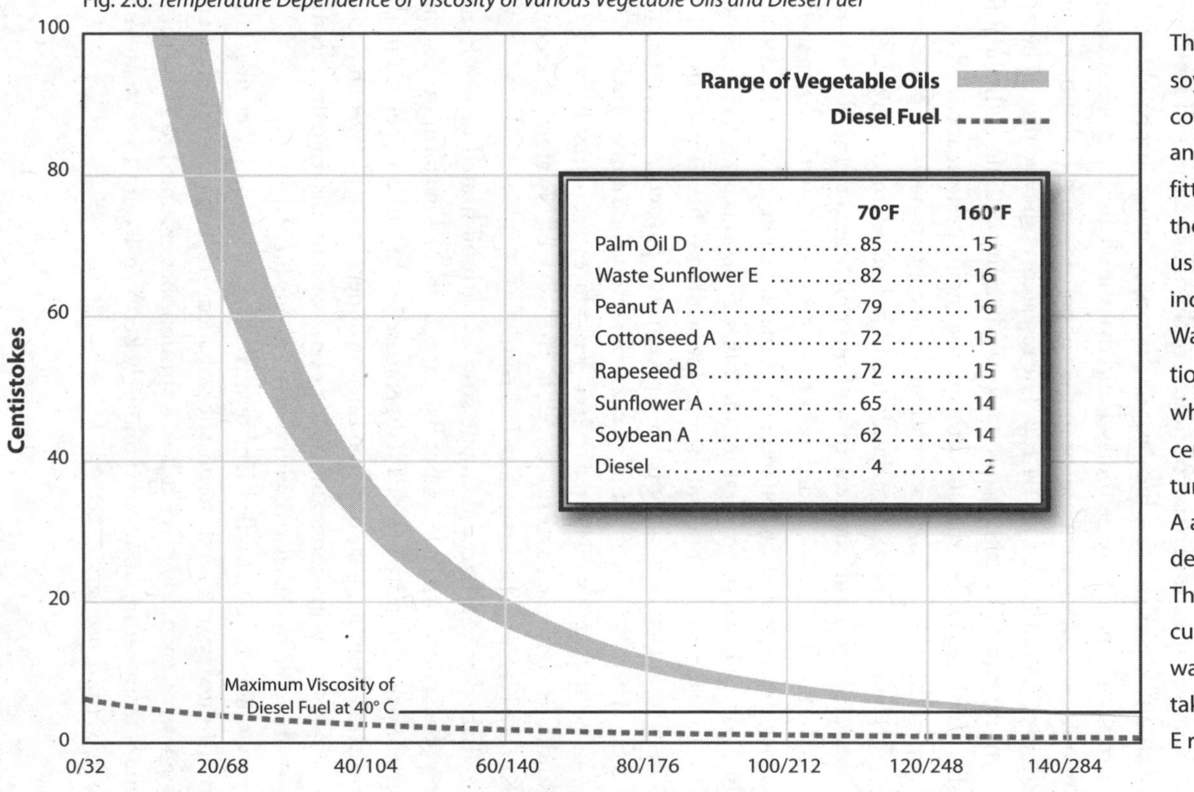

	70°F	160°F
Palm Oil D	85	15
Waste Sunflower E	82	16
Peanut A	79	16
Cottonseed A	72	15
Rapeseed B	72	15
Sunflower A	65	14
Soybean A	62	14
Diesel	4	2

The viscosity curves for soybean, sunflower, cottonseed, rapeseed, and peanut oil were fitted by the author with the software Gnuplot using data from the indicated source and the Walther-MacCoull equation, $v = 10^{T^{a}10^{b}} - 0.08$, where v is viscosity in centistokes, T is temperature in Kelvins, and B and A are experimentally determined constants. The formulas for the curves for canola and waste sunflower oil were taken from articles C and E respectively.[30a]

However, it is difficult to reach those temperatures in real world applications, and the literature suggests heating to 160° F is acceptable.

In an US Department of Agriculture (USDA) sponsored research program to develop a fuel specification for vegetable oil, Ryan, Dodge, and Callahan recommended that vegetable oil should be heated to 285° F before injection into the combustion chamber. They based the recommendation on their observation that oil heated to this temperature had a viscosity of around 4 centistokes, which was in the range, though just barely, of the viscosity specification of petrodiesel, and that the oil displayed injection and atomization patterns equivalent to that of petrodiesel fuel that had a viscosity of 2.4 centistokes at 104°F.[30] The researchers showed that oil heated to the recommended temperature combusted much more efficiently and fully than unheated oil, but did not compare the engine performance and emissions of the oil heated to 285° F with oil heated, but to a lower temperature.

In engine studies by Pugazhvadivu and Jeyachandran, these researchers found that fueling an engine with sunflower oil heated to 158°F compared to unheated, 86°F oil provided marked improvements in the thermal efficiency, production of carbon monoxide, fuel consumption, smoke density, increases in exhaust temperature and NO_X emissions, all indications of a higher completeness of combustion. Crucially, further heating the oil to 275°F produced only marginal improvements, particularly in a slightly higher thermal efficiency and lower carbon monoxide emissions.[31]

The finding that oil heated to 160° F produces a good results in engine studies is confirmed and supported in the work of Nwafor, [32,33] Senthil Kumar, Kerihuel, Bellettre, and Tazerour.[34]

Labeckas and Bari did engine studies with oil heated to 140°F, and reported good initial results; but Bari later found in long term studies that crude palm oil heated to this temperature led to carbon deposits in the injection chamber, wear of piston rings, cylinders liners, plunger and delivery valves of the injection pump, and poor spray patterns from the injectors — all problems associated with overly viscous fuel.[35]

Until we have controlled, long term, destructive studies of engines running heated vegetable oil, the question of adequate temperature will remain somewhat open. However, it is our opinion that around 158°F is a minimum temperature, and correspondingly between 13 and 15 centistokes is a maximum viscosity.

Convergence of Viscosity

As temperatures increase, the viscosities of various vegetable not only decrease but the differences in viscosities of different oils also narrow considerably (see Fig. 2.6). This is what allows us to set a minimum injection temperature for all vegetable oils, instead of having to have to specify a different injection temperature for every oil, and this is why, in a system with adequate heating, the viscosity of the fuel at ambient temperature really does not matter. In a system that heats the oil to 160°F by the time the fuel gets to the injectors, the oil will be thin enough to burn efficiently, even if the oil was like margarine when it was put in the tank.

Delay in the Temperature/Viscosity Relationship

In many applications, the oil is only heated before it enters the injection pump or high pressure fuel system, components that have a lot of thermal mass and which can affect the temperature of the oil. The amount of heat exchange that occurs in the high pressure fuel system, and whether the oil will be heated above or cooled from coolant temperature depends upon the particular engine, however, heating the oil as much as possible prior to its reaching the high pressure fuel system is never a wasted effort.

Despite claims to the contrary, there is no engine that I know of in which the amount of heat exchange in the high pressure system is so great that heating the oil prior to its entering the system does not significantly affect the temperature of the oil exiting the high pressure system. And, even in the engines with high pressure fuel systems that rob heat from the oil, the temperature drop is only a fraction of the temperature gain produced by earlier heating efforts. It may also turn out that small, quick drops in temperature have less impact on viscosity than we might imagine.

The viscosity of vegetable oil exhibits a property called hysteresis, which means that there is a time delay between changes in temperature and changes in viscosity.[36] When vegetable oil is heated from, say, 100° F to 140° F, the viscosity drops from 40 centistokes to 20. However, if we quickly cool that oil back down to 100° F, the viscosity of the oil does not immediately return to 40 centistokes. If we measure the viscosity within a few seconds the viscosity will be more like 30 or 35 centistokes. Of course a drop of 40° is just an example; such a large drop would be unacceptable and would require modifications to the fuel system. Whenever possible, temperature drops of any sort should be avoided.

Science — For Those Who Have to Know More

No, Really, What is Viscosity?

Viscosity is the resistance a liquid has to being deformed. What provides that resistance is the attraction that exists between the molecules of the liquid. The triglyceride is a nonpolar molecule. That means that there are no regions that are permanently charged positively or negatively. Water is a good example of the other type of molecule, a polar molecule. In a water molecule, the regions around the hydrogen molecules are always positively charged and the region around the oxygen molecule is always negatively charged. The attraction between oppositely charged "poles" of H_2O molecules make up most of the intermolecular attractive force in water (Fig. 2.7).

Even though a triglyceride molecule does not have permanent poles, the distribution of charge in the cloud of electrons that surround the molecule is constantly and randomly fluctuating. At any instance, one part of the molecule will be more negative and correspondingly another part will be more positive. In other words, negative and positive poles are always forming and disappearing. These poles can attract the randomly appearing

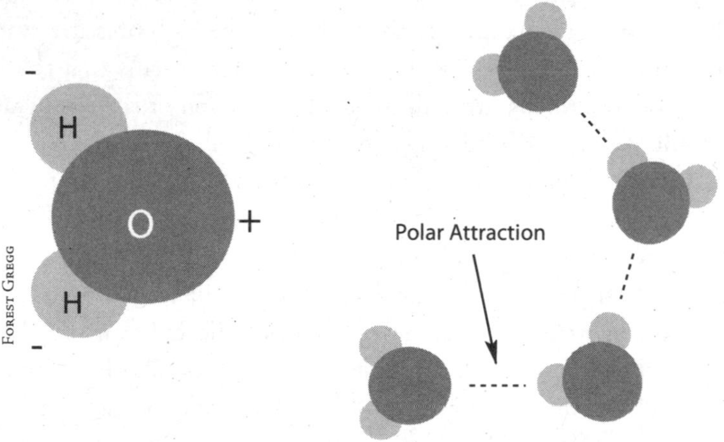

Fig. 2.7: *The attraction between oppositely charged "poles" of H_20 molecules make up most of the intermolecular attractive force in water.*

poles of neighboring molecules, or more stably, a pole can induce the formation of an oppositely charged pole in a neighboring molecule and develop a more stable attraction.

This kind of attraction is called London dispersion forces. London forces increase with the increases in the size of the molecule, because the larger the molecule, the more electrons in the cloud and the larger the potential for imbalance of charges. Let me say that again, the larger the molecules the greater the viscosity. This is why vegetable oil is more viscous than biodiesel and petrodiesel, because the average molecule of vegetable oil is three times larger. Among vegetable oils, this is why oils that contain more and longer average fatty acid chains are more viscous, all else being equal.

Unsaturation

The shape of the triglyceride also matters because two molecules of a regular and flexible shape can more easily align and develop more sites of attraction than molecules of more fixed and irregular geometry.

Imagine a straight, willowy reed. The reed has a lot of freedom to bend and can make a great number of shapes. Now imagine that the reed contains a hard, inflexible knot halfway down its length. This knot severely limits that shapes the reed can make (Fig. 2.8).

A double bond in the carbon chain of a fatty acid acts in a very similar way that the inflexible knot does in our reed. It prevents that fatty acid from easily conforming to the shapes of neighboring molecules, and for this reason unsaturated triglycerides cannot pack together as easily as saturated triglycerides and are thereby much less effective at forming the intermolecular bonds that make a liquid viscous.

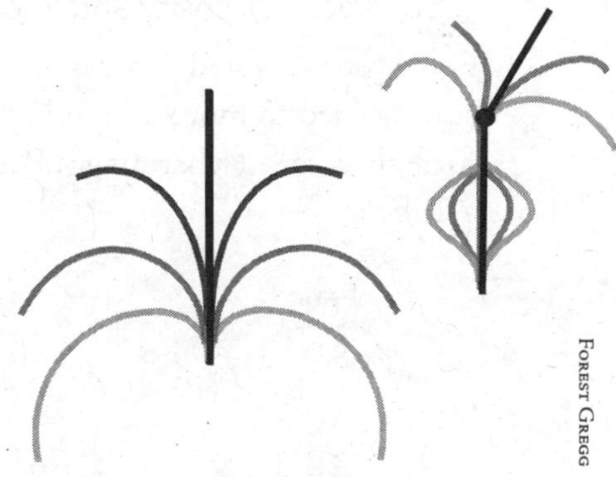

FOREST GREGG

Fig. 2.8: *Two molecules of a regular and flexible shape can align and develop more sites of attraction than molecules of fixed and irregular geometry. For example, a straight, willowy reed has a lot of freedom to bend and can take a great number of shapes. If the reed contains a hard knot halfway down its length, flexibility becomes limited.*

Chain length and degree of unsaturation determine viscosity at a given temperature, and many attempts have been made to develop accurate formulas to predict viscosity based upon those properties. One of the few that gives the predicted viscosity in centistokes is Goering, et. al. who came up with the following equation for the viscosity of vegetable oils at 100° F.[37]

Viscosity, in centistokes
= 73.14 + 6.409 *(Average Number of carbon Atoms)*
– 3.028 *(Average Number of Double Bonds)*^{2}

Of Poise and Centistokes

The differences between poise and centistokes are left to the reader to investigate, and a wealth of information can be easily found on the Internet. However, the reader may be in a position to want to convert between poise and centistokes. This is easily done using the formula:

Stokes = Poise/Specific Gravity

More About Viscosity and Fuel Line Filters

For those interested, there is actually a formula for figuring out the pressure needed to move a liquid of certain viscosity through a tube. This formula is called, charmingly, Pouisson's "Law of Fluid Moving Through a Tube":

$$v = \frac{\pi r^4 p}{8cl}$$

or

$$p = \frac{8vcl}{r^4}$$

v is volume in cm^3,
r is radius of the tube in cm,
p is pressure differential at the ends of the tube in $\frac{dynes}{cm^2}$,
c is viscosity in poise, and
l is length of the tube.

While I do not imagine that readers will rush out and measure all the radii, lengths, and viscosities of their fuel systems, this formula does bear examination. Notice that all the variables are linear except the radius, which is raised to the fourth power. That means that small changes in the radius will have outsized effects on the amount of pressure needed to move a given amount of fuel through the system, a strong argument for making fuel lines and filters as large as practical.

Ignition Delay

The ignition delay is the time that passes from when a fuel is injected into the combustion chamber until it autoignites. Ignition delay affects everything: knock, noise, power, efficiency of combustion, and emissions. The standard indicator of ignition delay in diesel fuels is the cetane number, where a larger cetane number means a shorter ignition delay.

Petrodiesel is specified to have a cetane number of at least 40. The measured cetane number of vegetable oil ranges between 30 and 50, but all the cooking oils common in this country have cetane numbers below 40 (see Table 2.3), indicating that these oils should have a longer ignition delay than petrodiesel, and should be unacceptable substitutes for diesel fuel.

	Cetane Number	Iodine Value
Linseed	27.6	156.74
Bay	33.6	105.15
Walnut	33.6	135.24
Cottonseed	33.7	113.2
Almond	34.5	102.35
Peanut	34.6	119.55
Wheat	35.2	120.96
Poppy	36.7	116.83
Sunflower	36.7	132.32
Rapeseed	37.5	108.05
Corn	37.5	119.41
Soybean	38.1	*
Sesame	40.4	91.76
Safflower	42.0	139.83
Castor	42.3	88.72
Olive	49.3	100.16
Hazelnut	52.9	98.62

Source:
A. Demirbas, "Chemical and fuel properties of seventeen vegetable oils," *Energy Sources* 25, no. 7 (2003): 721-728, http://dx.doi.org/10.1080/00908310303400

Table 2.3 : *Chemical and Fuel Properties of Seventeen Vegetable Oils.*

It turns out, however, that cetane number is a poor indicator of the actual ignition delay of heated vegetable oils. In engine studies, heated vegetable oil consistently has a shorter ignition delay than indicated by the measured cetane number of the oil, and indeed a shorter ignition delay than petrodiesel.

Unfortunately, there is no other accepted measure besides cetane number that would allow us to quantitatively compare the ignition delay of heated oil and petrodiesel, and the best we can do is to say that common cooking oils, when heated above 160°F have relatively short ignition delays that make them acceptable as substitute diesel fuels.

Implication of Short Ignition Delay of Heated Vegetable Oil

Because of a shorter ignition delay, heated vegetable oil tends to be more thermally efficient, emit less carbon monoxide and hydrocarbons, and produce more NO_X than petrodiesel.

Thermal Efficiency. Thermal efficiency is how much of the heat energy stored in the molecules of the fuel is transformed into useful, mechanical work. In this case how much of the heat energy in the vegetable oil molecules is put to work pushing the piston. Engine studies of heated vegetable oil consistently show a slight improvement of thermal efficiency over petrodiesel.[38]

Effects of Ignition Delay

Ignition delay is important because it controls how explosively diesel fuel combusts. If a fuel has a long ignition delay, then most or all of the fuel will be injected into the combustion chamber and mixed with air before ignition begins, and once it begins all the fuel will burn at once, and produce much higher peak pressure and peak temperatures. With a shorter ignition delay, the combustion begins while there is still relatively little fuel in the chamber and continues longer at less intensity.

Think about a cigarette lighter. If you released the gas into an empty beer mug and struck the flint you would get a sudden intense flame. But release the same amount of gas into a burning lighter flame you get a continuous, lower intensity flame — the principle is the same in the combustion chamber. Within limits, a shorter ignition delay means a slower, less intense, and longer combustion.

It may seem like a more intense explosion would be desirable since it is the force of the explosion that powers internal combustion motors, but it turns out that it is mechanically more efficient to have a less intense pressure increase that pushes longer on the piston than a more intense pressure increase that is very short. Furthermore, a longer combustion period also produces a more even and more efficient burn. There are fewer cool pockets of mixtures of fuel and air, which in turn means more complete combustion, meaning that more of the fuel injected into the chamber is getting put to use, along with less uncombusted fuel and carbon monoxide being pumped into the air.

You may be wondering at this point if a short ignition delay has so many benefits, why doesn't diesel fuel have a higher cetane rating than it does? The main reason is NO_X, an EPA-regulated pollutant implicated in the formation of smog and acid rain. NO_X is produced by the high temperatures of combustion, and a longer period of combustion created by a shorter ignition delay means a longer period of time when conditions are right for NO_X to form.

Ignition Delay and Unsaturation

Vegetable oils that are more unsaturated tend to have longer ignition delays. Though cetane numbers do not accurately capture the ignition delay of heated vegetable oil, the difference between cetane numbers of different vegetable oils does reflect differences in the length of ignition delay between those heated oils in actual engines.[39] (Fig. 2.9). The cetane-lowering effect of unsaturation is explained by the greater susceptibility of unsaturated oils

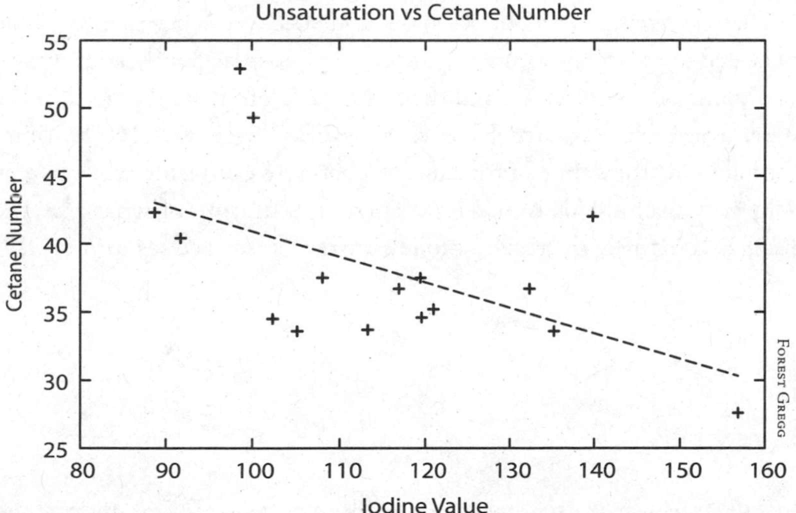

Fig. 2.9: *Vegetable oils that are more unsaturated tend to have longer ignition delays. The difference between cetane numbers of different vegetable oils reflects differences in the length of ignition delay between those heated oils in actual engines.*

to chemical degradation during precombustion, a topic that will be dealt with in the section on high-temperature reactions on the next page.

Advanced Topic

What is Cetane? Cetane number is really a measure of how similar a diesel fuel is to cetane, a hydrocarbon that has a very short ignition delay, under very specific test conditions. Using a Waukesha Cetane Test Engine, and following a defined testing protocol, the ignition delay of a test diesel fuel is compared with the ignition delay of a mixtures of cetane and isocetane, another hydrocarbon with a very long ignition delay. Percentages of cetane vs isocetane are adjusted until the mixture has the same ignition delay as the test fuel; the percentage of cetane in the matching mixture is the cetane number. ASTM-certified petrodiesel fuel has to have a cetane number above 40, which means it has to have the same ignition delay as a mixture of cetane and isocetane that is more than 40 percent cetane (Fig. 2.10).

If the researcher deviates from the protocol, the result is no longer a valid cetane number. This is why cetane number, as it is defined currently, cannot accurately capture the ignition properties of heated vegetable oil. The test that defines a valid cetane number specifies that the test fuel must be heated and held at a temperature between 65° and 90°F. We know that vegetable oil heated to that temperature will be too viscous, spray poorly, and atomize slowly out of the injectors, and the low cetane numbers reported in the literature are to be expected. However, if we heat the oil up to 160°F , we are not following the defined protocol, and therefore our result, while potentially very useful, fails to meet the current definition of what a cetane number is (namely the result of following the prescribed test protocol).[40]

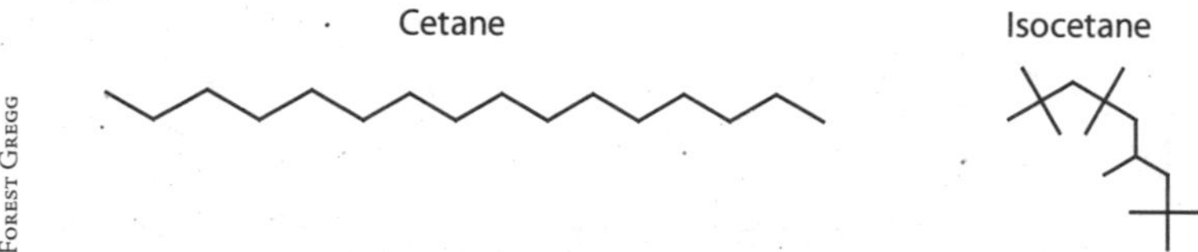

Fig. 2.10: *Cetane number is a measure of how similar a diesel fuel is to cetane, a hydrocarbon that has a very short ignition delay, under very specific test conditions. Certified petrodiesel fuel must have a cetane number above 40, which means it has to have the same ignition delay as a mixture of cetane and isocetane that is more than 40 percent cetane.*

High Temperature Reactions: Life in the Combustion Chamber

As we've alluded to a number of times, when vegetable oil fuel first enters the combustion chamber it undergoes a number of chemical changes that affect how the oil will combust and burn.

Thermal Cracking and Polymerization

Surprising Results. Research on petroleum-derived fuels showed that increasing the viscosity of the fuel narrowed the angle of the spray cone and increased the penetration rate, i.e. the speed at which the spray moved away from the injector.[40a]

However, when researchers studied the spray patterns of vegetable oil they found something very surprising. As they thinned the oil by heating it up, the spray pattern did the opposite of what they expected. The spray of hotter, thinner oil was narrower and had a higher penetration rate than colder, thicker oil. This pattern held until the oil was heated to all the way up to 293° F, the maximum test temperature (Fig. 2.11).[40.b]

Also surprising, the researchers found that the spray pattern of vegetable oil was nearly identical to petrodiesel when the viscosity of the oil was still

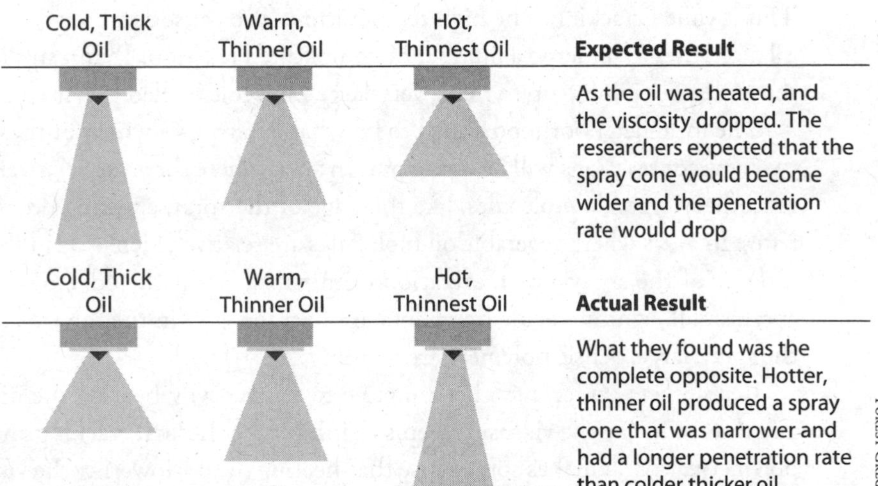

Fig. 2.11: *The spray patterns of petroleum-derived fuels and vegetable oil differed unexpectedly with a change in viscosity in research trials.*

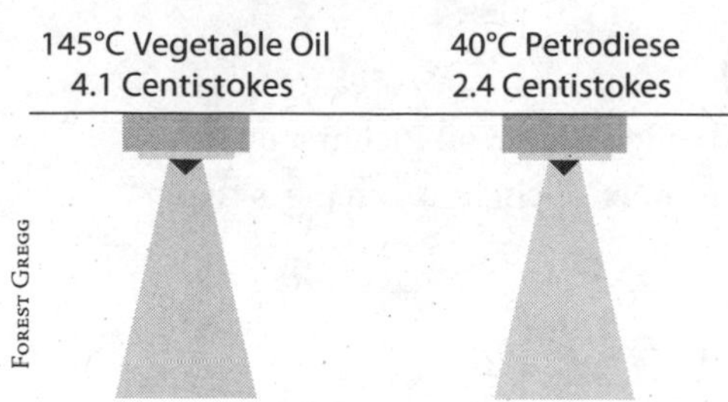

Fig. 2.12: *The spray pattern of vegetable oil was nearly identical to petrodiesel with the viscosity of the oil still nearly double that of the conventional fuel. Oil heated to 285°F with a viscosity of 4.1 centistokes looked the same as the spray pattern of unheated petrodiesel that had a viscosity of 2.4 centistokes at 104°F.*

nearly double that of the conventional fuel. Specifically, oil heated to 285° F with a viscosity of 4.1 centistokes looked the same as the spray pattern of unheated petrodiesel that had a viscosity of 2.4 centistokes at 40° C (Fig. 2.12).

The researchers explained these surprising results by figuring that something must be happening to the oil, specifically thermal cracking and polymerization.

The Mechanism. The combustion chamber is a hot place. When the injectors spray the fuel into the chamber, the temperature of the chamber can be in excess of 1,000° F. At these high temperatures, vegetable oil apparently does two contradictory things at the same time: it breaks apart and it clumps together. The high temperature causes some vegetable oil molecules to spontaneously break apart into two or more smaller fragments. This is called cracking. The high temperatures also cause some vegetable oil molecules to undergo a number of complicated reactions that result in the molecules joining up to make very large molecules called polymers.

The local neighborhood seems to be what determines which of these two types of reactions will be dominant. In areas where there are relatively few other vegetable molecules, like the edge of the spray, cracking dominates. In areas where vegetable oil molecules are relatively dense, as in the interior of the spray, polymerization dominates. So, at the edge of the spray, small, light molecule fragments are forming, and in the interior of the spray, large, dense, polymers are forming.

Researchers, so far, have been unable to answer why heating the oil, and thus reducing the viscosity, seems to inhibit the thermal cracking and polymerization. It makes some sense that heating the oil, lowering the viscosity, and thereby improving atomization, would inhibit vegetable oil from polymerizing, as that is a process that requires the fuel to be liquid, and better atomization would shorten the period until all the fuel evaporated.

But it is not at all clear why heating the oil or reducing the viscosity would inhibit thermal cracking. However, as they investigated the phenomena, the researchers did explain the connection between higher unsaturation and ignition delay.

Unsaturation and Ignition Delay. More unsaturated oils tend to have longer ignition delays and higher cetane numbers,[41,42] and the reason for this is that more unsaturated oils break down more in the period between when the fuel is injected and when it autoignites, and produces more ignition-delay-lengthening compounds.

As we've already discussed, unsaturated fatty acids are more unstable than their saturated cousins. This is true for oxidative polymerization at normal temperatures and it is also apparently true for the complex reactions occurring at the very temperatures of the combustion chamber.

Other Fuel Properties

Specific gravity

Specific gravity is a measure of how dense a substance is. At 68°F, petrodiesel has a specific gravity of 6.83 pounds per gallon. Vegetable oils are denser with specific gravities between 7.51 pounds per gallon and 7.59 pounds per gallon, again at 68°F. This means that if vegetable oil and diesel are not dissolved into each other, that diesel will float on top of vegetable oil. Water has a specific gravity of 8.35 pounds per gallon at 68°F. Both diesel and vegetable oil float on top of water, but the smaller difference between the density of vegetable oil and water compared to the density of diesel and water partially explains why water separators designed for diesel fuel do not work as effectively with vegetable oil.

Bulk Modulus

Bulk modulus is a measure of how much pressure is necessary to compress a liquid by a certain amount. Vegetable oil is less compressible than petrodiesel with a bulk modulus of about 275,000 psi compared to about petrodiesel's 232,000 psi.[43] The bulk modulus is important because, together with specific gravity, it determines the speed at which a pressure wave moves through the fluid. The equation for this is:

When you solve for this equation, you'll find that pressure waves move somewhat faster through vegetable oil than through petrodiesel. This means that when vegetable oil is the fuel, the pressure wave created by the injection pump moves more quickly through the injection lines to the injector, which has the affect of slightly advancing the timing.

Bulk Modulus

Bulk modulus is a measure of how much pressure is necessary to compress a liquid by a certain amount. Unheated vegetable oil, with a bulk modulus of about 275,000 psi, is less compressible than petrodiesel's bulk modulus of 232,000 psi.[i]

The bulk modulus is important because it determines two factors that are critical for diesel engines with injection pumps: pressure, and the speed that a pressure wave moves through a fluid. In these types of systems, the higher bulk modulus causes the injectors to open and close later, inject vegetable oil at a higher pressure, and this leads to unintended secondary injection and nozzle dribble. As vegetable oil is heated, the bulk modulus drops, and these problems lessen or disappear completely.

Functioning of Injection Pumps

Both inline and rotary injection pumps work by compressing fluid in a cylinder in the pump. This compression creates a pressure wave that travels from the pump, through the injection lines, to the injectors. Once the pressure of the fuel in the injectors is high enough to overcome a spring-loaded seat, the fuel then sprays into the combustion chamber. That opening pressure is called the pop pressure. When the fuel pressure drops below the spring pressure of the injector, the injector closes and the spray stops. That pressure is called the closing pressure.

Interaction of Bulk Modulus

Because vegetable oil has a higher bulk modulus than diesel fuel, when it's compressed by the injection pump, it develops a higher fuel pressure. This alone would cause the vegetable oil fuel to reach pop pressure earlier and stay above high pressure longer than diesel fuel, leading to earlier injection and later end of injection.

However, bulk modulus also controls the speed of the pressure wave moving through the fuel, and the higher the bulk modulus the faster the compression wave travels. ☞

Specific Heat

Specific heat is the amount of energy needed to raise the temperature of a given amount of a substance one degree, most commonly measured in calories. The specific heat of vegetable oil is less than half of the specific heat of water. Practically, this means that it is easier to increase the temperature

This means that not only is the opening pressure reached more quickly, because the fuel is more resistant to compression, but that the pressure wave that moves from the injection pump to the injector is also moving faster. This also has the affect of advancing the beginning of fuel injection.

The story is made even more complicated by the fact that pressure waves, like all waves, bounce and refract when they hit surfaces and interact with each other in complicated ways. After the point that the injector is supposed to close, there are still pressure waves moving back and forth in the injection line that can cause the injector to open again, or fail from closing tightly, a phenomena called secondary injection, or nozzle dribble, respectively.

This failure to have a clear end of injection leads to serious problems with carbon deposits on the injector.[ii] The fuel that comes out after the primary injection is entering a combustion chamber that has already become too cool for quick combustion and it is very poorly dispersed, ensuring that there will be local areas where there is not enough oxygen to burn the fuel — a perfect recipe for the formation of carbon deposits.

Bulk Modulus and Temperature

Fortunately, bulk modulus drops with increases in temperature. By the time it is heated to 160° F, vegetable oil has the same bulk modulus as room-temperature petrodiesel.[iii] ■

i. Rakopoulos, C.D., K.A. Antonopoulos, and D.C. Rakopoulos, "Multi-zone modeling of Diesel engine fuel spray development with vegetable oil, bio-diesel or Diesel fuels," *Energy Conversion and Management* 47, no. 11-12 (July 2006): 1550-1573, (accessed September 24, 2007).

ii. Baranescu, R.A., and J.J. Lusco, "Performance Durability and Low Temperature Evaluation of Sunflower Oil as a Diesel Extender," in *Vegetable Oil Fuels: Proceedings of the International Conference on Plant and Vegetable Oils as Fuels* (St. Joseph, MI: American Society of Agricultural Engineers, 1982), 312-328

iii. Varde, K.S., "Bulk Modulus of Vegetable Oil-Diesel Fuel Blends," *Fuel* 63, no. 5 (May 1984), doi:10.1016/0016-2361(84)90172-8 of oil than it is to increase the temperature of water. Specific heat mainly comes into play when designing the systems to heat vegetable oil to an appropriate fuel temperature.

For those who need it, at 68° F, the specific heat of vegetable oil ranges from 1.83 to 2.20 $J{\cdot}kg^{-1}{\cdot}C^{-1}$, and water has a specific heat of 4.18 $J{\cdot}kg^{-1}{\cdot}C^{-1}$.

Lubricity

Vegetable oil is an excellent lubricant, much better than petrodiesel. Lubricity is often thought of as determined solely by viscosity, with thicker fluids being better lubricants. However, in the case of vegetable oil, while viscosity affects the lubricity of vegetable oil it does not explain it. In other words, the lubricity is largely determined by a poorly understood collection of chemical and physical attributes of oil.[44]

Heating diesel fuel should be avoided, because what little lubricating property petrodiesel has is diminished by heat. However, there is no such concern for vegetable oil.

Solvency

In older cars, using certain biofuels will cause serious damage to the fuel system as the biofuels are often much more powerful solvents than the conventional fuel. Biodiesel, for example, will dissolve the seals and flexible fuel lines in many vehicles made before 1993. Volkwagens in particular are at great risk, as biodiesel dissolves seals in the injection pump and biodiesel will then drip on and dissolve the rubbery timing chain. When that fails, the pistons will hit the valves and you will be lucky if all you have to do is replace the top end of the engine. Diesel vehicles made since 1993 usually have seals and flexible fuel lines that are biodiesel-compatible.

Vegetable oil does not have nearly the solvent strength of biodiesel, or diesel for that matter, so it does not lead to the solvency related problems that are associated with most biofuels. Below is a table of material compatibilities with vegetable oil, diesel fuel, and biodiesel (see Table 2.4). This is a very general guide, and should not be the final basis for material selection. The behavior of these materials can vary significantly depending upon the variety of material. Further, temperature and pressure can powerfully affect the interactions between material and fuel. Hard data is difficult to come by, but for those who need to know, the Hildebrand Solubility Parameters for petrodiesel is less than 15[45], for vegetable oil about 7-8[46], and for biodiesel about 16-17[47,48].

Cold Flow Properties

In cool weather, both vegetable oil and diesel begin to gel and form solids that can clog filters and eventually turn the fuels into a solid jelly-like mass that won't flow. The petroleum industry uses four measures to describe the cold temperature characteristics of fuel, two of which are easily accessible: and pour point cloud point.

In petrodiesel, as temperatures fall, the first solids to form will be large molecules called parafinnic waxes coming out solution, similarly to how salt will fall out solution when hot, very salty water is cooled. If the

Material	Vegetable Oil	Petrodiesel	Biodiesel
PVC	A	B	A
Tygon	B	A	D
Nylon	A	A	B
High Density Polyethylene	A	B	B
Cross-linked Polyethylene (PEX)	B	D	B
Polypropylene	A	A	D
Polyurethane	B	C	B
Silicone	A	D	B
Neoprene	C	B	D
Nitrile (Buna N)	A	A	D
Viton	A	A	B
Natural Rubber	D	D	D
Fluorocarbon	A	A	A
Teflon	A	A	A

A: Excellent, **B**: Good, **C**: Fair, **D**: Unacceptable
Sources: "Cole-Parmer: Chemical Resistance Database," http://www.coleparmer.com/techinfo/chemcomp.asp
National Biodiesel Board, Materials Compatibility, http://www.biodiesel.org/pdf_files/fuelfactsheets/Materials_Compatibility.pdf
McMaster-Carr, "Chemical Compatibility of EVA and Polyethylene Tubing," http://www.mcmaster.com/addlcontent/loadaddcontent.asp?doc=9334TAC

Table 2.4: *Compatibility of Various Materials with Vegetable Oil, Petrodiesel, and Biodiesel.*

temperature continues to drop, liquid portions of the fuel will begin to freeze, eventually forming an unpourable gel. The temperature at which this happens is the fuel's *pour point*. For vegetable oil, the solids are mostly formed through this second mechanism. As the temperature falls, high-melting-point trigycerides begin to freeze and form large fat crystals.

The cloud point is the temperature at which the clear fuel first becomes hazy. It is an easy change to measure, but a fuel that has just reached its cloud point will usually pass through filters and the fuel system without problems. The industry has a measure called the cold filter clogging point, that better measures what we care about — when the fuel is too cold to pass through a filter — but it requires testing apparatus that most people will not be interested in acquiring. For most of us, the cloud point is an acceptable indicator of when we need to start being worried about filter clogging.

Diesel fuel doesn't have any general specified cloud point or pour point. Instead, the refineries are required to make blends of fuel tailored for different geographical areas and the time of year. This usually works out well, except when there is a cold snap.

Vegetable oils vary widely in their cloud and pour points, due to differences in typical fatty acid profiles (see Table 2.5). Generally, more saturated oils have higher cloud and pour points than unsaturated oils. Not reflected in the table, used cooking oil usually has a higher cloud and pour point than fresh oil.

Flash Point and Autoignition Point

The flash point and autoignition points of vegetable oil are mainly of interest in regards to safety and handling. They are of limited use in understanding how the fuel behaves under the high-pressure conditions of a combustion chamber.

Flash Point. The flash point is the lowest temperature at which a fuel can form an ignitable mixture of fuel vapor and air at atmospheric pressure. If the ignition source is removed, fuel at the flash point will likely cease burning. Common vegetable oils have a flash point between 320°F and 610°F.[49] Petrodiesel has a flash point between 100°F and 130°F.[50]

Autoignition point. The autoigntion point is the temperature at which a fuel will produce a vapor that will spontaneously combust in the presence of air, at atmospheric pressure. Common vegetable oils have an autoignition point between 575°F and 850°F [51]. Petrodiesel has an autoignition temperature between 490°F and 545°F [52].

Stoichiometric Air/Fuel Ratio

The stoichiometric air/fuel ratio is the ratio of the mass of air to mass of fuel where, at the end of combustion, there would be no oxygen left over in the air and there would be no unburned or partially burned fuel. The stoichiometric ratio for diesel fuel is around 15:1, for vegetable oil it is about 13:1.[53]

Under typical conditions, diesel engines run lean, which means that there is more air in the combustion chamber than is necessary for complete combustion. Black smoke when accelerating is usually the result of overfueling past the stoichiometric ratio. More fuel is being put into the combustion chamber than can be combined with the oxygen in the air, and the unburned portion forms soot and makes black smoke.

Chemical Breakdown of Vegetable Oil

Compared to petrodiesel, vegetable oil is a relatively unstable fuel. If exposed to air, sunlight, high heat,

	Celsius		Fahrenheit	
	Cloud Point	Pour Point	Cloud Point	Pour Point
Corn	-1	-40	30	-40
Rapeseed	-4	-32	25	-25
H.O. Safflower	-12	-21	10	-5
Cottonseed	2	-15	35	5
Linseed	2	-15	35	5
Sunflower	7	-15	45	5
Corn	-1	-40	30	-40
Rapeseed	-4	-32	25	-25
H.O. Safflower	-12.	-21	10	-5
Cottonseed	2	-15	35	5
Linseed	2	-15	35	5
Sunflower	7	-15	45	5
Soya Bean	-4	-12	25	10
Crambe	10	-12	50	10
Sesame	-4	-9	25	15
Peanut	13	-7	55	20
Safflower	18	-7	65	20
Babassu	20	?	68	?
Palm	31	?	88	?
Light Hydrogenated Soybean*	32-37	2	90-100	36

Source: A. Srivastava and R. Prasad, "Triglycerides-based diesel fuels," *Renewable and Sustainable Energy Reviews* 4, no. 2 (June 2000): 111-133, http://www.sciencedirect.com / science/article/B6VMY-3YF46C4-1/2/ 2c56f49bb2482c2160aa2a5c5962b60e

* T.W. Ryan, L.G. Dodge, and T.J. Callahan, "The effects of vegetable oil properties on injection and combustion in two different diesel engines," *Journal of the American Oil Chemists' Society* 61, no. 10 (October 5, 1984): 1610-1619, http://dx.doi.org/10.1007/BF02541645

Table 2.5: *Cloud Point and Pour Point of Various Oils Used as Fuel.*

steam, or enzymes, vegetable oil will break down, and eventually become unusable as fuel. When excessively deteriorated, vegetable oil can cause problems such as shorter fuel filter life, paint-like or gummy deposits in the tank and fuel system, failure of the injection pump and injectors, contamination of the engine oil, and related catastrophic engine failure. From polls we've conducted and what we've seen in the shop, I'd say that between a quarter and a third of the vegetable oil conversions have suffered from fuel that had gone bad at some point, usually to no lasting ill effect.

The remainder of this chapter will go into some detail about one of the mechanisms that causes vegetable oil to go bad, and like anything truly interesting, they can be complicated, so let me just lay out the take-home points: Avoid oil from an open dumpster, don't pick up oil that would smell offensive to a maggot, store oil in full, preferably nonmetallic containers in a cool dark place, avoid exposing oil to metals, particularly copper and steel, and size your fuel tank so that a full tank is consumed in no more than a week, on average.

Oxidative Polymerization

Oil paints have traditionally had vegetable oils as their base, usually linseed oil. Expose paint to air, and the vegetable oil base will turn from a liquid to a solid. It will cure or "dry." The chemical process that causes paint to dry is called oxidative polymerization, and the same process that turns liquid paint into a solid skin can happen, to a lesser degree, with other oils. In a vegetable oil fuel system, oxidative polymerization can increase the viscosity of the oil, clog filters, form paint-like skins and gummy deposits in the fuel system, and if the oil contaminates the lubricating oil, quickly turn the engine oil into a mayonnaise-like substance, leading to catastrophic engine failure (Fig. 2.13).

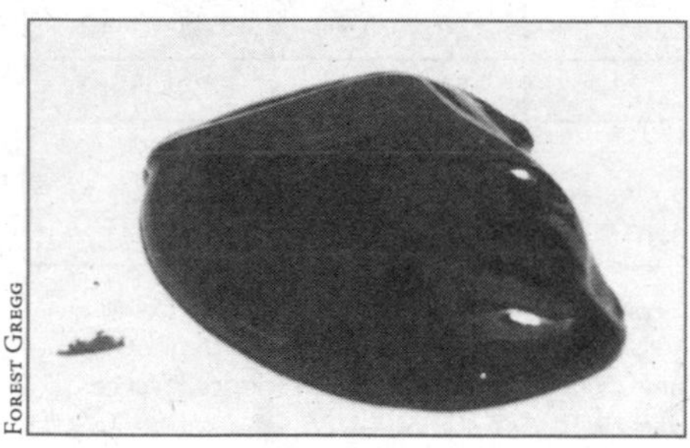

Forest Gregg

Fig. 2.13: *Lubricating oil contaminated with polymerizing vegetable oil. Oxidative polymerization can increase the viscosity of the oil, clog filters, and form gummy deposits in the fuel system. If the oil contaminates the lubricating oil, it can turn the engine oil into a mayonnaise-like substance.*

Whether or not oxidative polymerization will actually cause a problem in a specific conversion, depends upon number of factors: unsaturation of the the oil, presence of metal catalysts, temperature, exposure to air and sunlight, anti-oxidants, and time.

Factors of Rate of Oxidation

Unsaturation. Linseed, safflower, poppy, and walnut oils make good bases for oil paints because they are highly unsaturated, which means that they have many double carbon-carbon bonds, i.e. unsaturated bonds. These bonds are the reactive sites of oxidative polymerization, and the more of these bonds an oil has the more quickly and thoroughly it will break down. As a result, these oils will "dry" fairly quickly at room temperature.

Since unsaturation leads to chemical instability, it may seem like a good idea to use saturated oils, or oils with very few double bonds. Unfortunately, unsaturation is what makes vegetable oil liquid at room temperature. Highly saturated oils are no longer oils, but are fats, butters, lards, and tallows — solids or semisolids at room temperature. While an effective conversion can use solid fats as fuel, it is much more involved to pump and filter lard than oil, so most people avoid solid or semisolid fat.

Cooking oil in the US falls in the middle. The oil is unsaturated enough to be liquid at room temperature, but not so unsaturated that it will have a very short life in a fryer. The ideal oil would be one in which all the unsaturated fatty acids contained only one double bond. Unfortunately, the only oil that approaches that profile is olive oil, not the cheapest substance one could imagine.

Metals. Metals, even in small amounts will speed up oxidative polymerization. In particular, copper, iron in the form of mild steel, and brass, an alloy of copper and zinc, are common metals that should be avoided (Fig. 2.14). Aluminum is the least reactive common metal, and higher grades of stainless steel are acceptable. Table 2.6 shows the concentration of various metals that will cut the oxidative stability of an oil in half.

Metal	mg/kg
Copper	.5
Iron	.6
Manganese	.6
Chromium	1.2
Nickel	2.2
Vanadium	3.0
Zinc	19.6
Aluminum	50.0

Source: J.B. Rossell, ed., Frying - Improving Quality (Woodhead Publishing, 2001), http://www.knovel.com/knovel2/Toc.jsp?BookID=544&VerticalID=0 (accessed July 12, 2007).

Table 2.6: *Metal Concentration where Oxidative Stability of Oil is Halved.*

In addition to accelerating the reaction of oxidative polymerization, metals containing iron also seem to affect the end result. For reasons that are not well understood,

Fig. 2.14: *Metals tend to speed up oxidative polymerization. Copper, mild steel, and brass are common metals that should be avoided in vegetable oil systems.*

the "dry" paint-like skins seem more likely to form on metals that contain iron, particularly mild steel and, to a lesser extent, on lower grades of stainless steel.

Temperature. Edible vegetable oils will oxidize slowly below 70°F, but the reaction occurs quickly at temperature of 100°F or above. The rate of oxidation will double with every increase of 50°F.[54]

Storage. Vegetable oil can last for years if stored in closed, opaque containers at room temperature or cooler. If a container is closed, the supply of oxygen is cut off, and oxidative polymerization cannot proceed. Sunlight, or other sources of ultraviolet light, will rapidly degrade oil, and should be avoided. The clear plastic jugs that restaurants commonly buy their oil in are permeable to oxygen and should be avoided for long term storage.

Antioxidants. There are number of chemicals on or near market that will retard oxidative polymerization. Common examples are THBQ and BHT, which are commonly added to vegetable oil in the refinery to prolong shelf life. However, these and other common chemicals are not potent enough to offer real or economic protection for vegetable oil fuel. We at

Frybrid have been working on bringing an effective additive to market that was originally developed for the biodiesel industry, and I expect that this product or one very much like it will be available in the next few years.

Time. Like anything else, oxidative polymerization takes time. How much time is near impossible to say, as it depends so much on what the oil is, the history of the oil, the ambient temperature, the particulars of the conversion, how the oil was stored, and even one's driving habits. I can say that those who have reported problems tend to have not used their vegetable oil system for weeks on end, have been improperly storing their oil for prolonged lengths of time, or have vegetable oil fuel tanks that are too large for their driving patterns, e.g.: having a 90-gallon tank when they only use 15 gallons a week.

Telling if Vegetable Oil is Oxidized. There is unfortunately no simple objective test to tell if vegetable oil is too oxidized. However, you have one good, sophisticated tool to detect the oxidative breakdown of oil: your nose. At different stages of the oxidative breakdown of vegetable oil, different types of compounds are produced that have characteristic smells that can give you an indication of how far gone the oil is.

Table 2.7 below shows a typical sensory evaluation board, as might be used in a vegetable oil refinery. At the refinery, they have a group of folks

Flavor Score		Description of Flavor
10		Completely Bland
9	Good	Trace of flavor, but not recognizable
8		Nutty, sweet, bacony, buttery
7	Fair	Beany, hydrogenated
6		Raw, oxidized, musty, weedy, burnt, grassy
5	Poor	Reverted, rubbery, butter
4		Rancid, painty
3	Very poor	Fishy, buggy
2		Intensive flavor and objectionable
1	Repulsive	Wiley

Source: Francis, Frederick J. 1999. *Wiley Encyclopedia of Food Science and Technology* (2nd Edition). Volumes 1-4. John Wiley & Sons. Online version available at www.knovel.com

Table 2.7: *Odor Evaluation Board.*

who actually are paid to taste the oil. I'm not suggesting that. I thinking smelling it is just fine. While smells from the food can mask the odor of the oil itself, as the oil deteriorates the aroma will get quite strong.

In my experience, when the oil begins to smell grassy, I would have my doubts about using it as fuel, and when it begins to smell painty I know I have a problem.

Oxidation and Color. Another potentially useful indicator of oxidation is color. As vegetable oil oxidizes it will darken and then eventually lighten.[55] Many restaurants change the oil once it has darkened; if the oil in storage or in the tank has begun to lighten in color that is a fairly good indicator of advanced oxidation.

Oxidative Polymerization and Filtering. When your nose tells you that the vegetable oil is on its way out, there's really not much you can do with it as fuel for an engine. Filtering will *not* make oxidized oil safe to use. As illustrated in Fig. 2.15, filtering will only removes the larger oxidization products leaving the bulk of the reactive particles in the oil. Refer to the section on filtering in the Practical Concerns chapter for ideas about what to do with unusable vegetable oil.

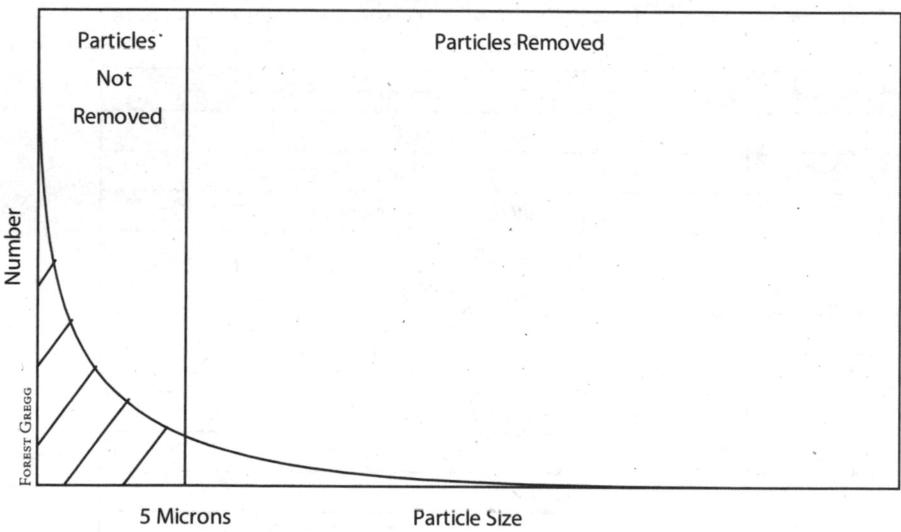

Fig. 2.15: *Filtering removes the larger oxidization products, but leaves the bulk of the reactive particles in the oil.*

The Chemistry of Oxidative Polymerization. Oxidation polymerization is a two-step process. First comes oxidation. Oxidation simply means the addition of oxygen to a substance. Fire, rust, and brown apples are all examples of oxidation. In this case, we are concerned with the oxidation of unsaturated fatty acids.

As you may remember from the chapter on triglycerides, most vegetable oil molecules contain at least one, bent, double carbon-carbon bond in a fatty acid. These double bonds, also known as unsaturated bonds, are what make vegetable oil susceptible to oxidative breakdown.

The unsaturated double bonds are not the directly involved in the chemical changes. Instead, the unsaturated bond causes the carbon atoms next to the bond to have weaker holds on their hydrogen atoms. From time to time, one of these relatively loosely held hydrogens will split off (Fig. 2.16)[56].

After losing the hydrogen, the carbon is now a very reactive radical, meaning an atom that wants very much to form a bond with another atom, and may even steal an atom from another molecule to do so.

Now, if the radical carbon reattaches with an atom of hydrogen, then we've come back to beginning and all is right in the world, for now.

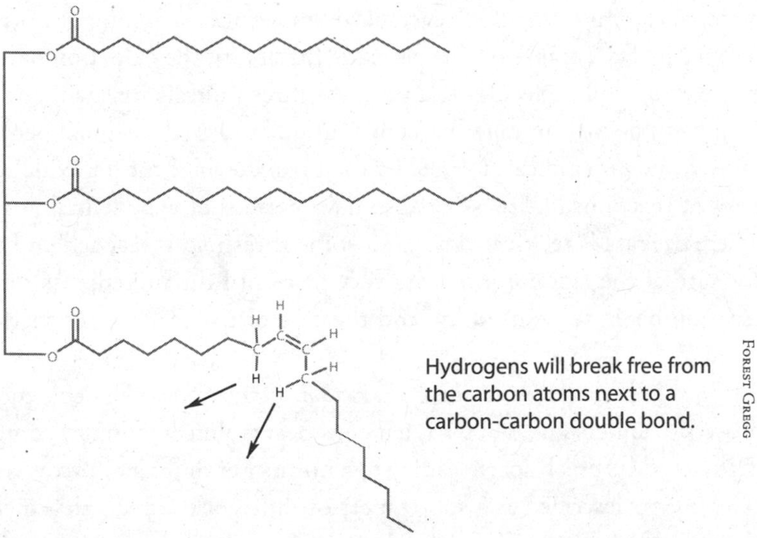

Fig. 2.16: *Loosely-held hydrogens will break free from the carbon atoms next to a carbon-carbon double bond.*

However, if the radical carbon comes into contact with a molecule of atmospheric oxygen, things become much more complicated.

The oxygen in the air is mainly in the form of O_2, two oxygen molecules sharing a double bond. Given the opportunity, our radical carbon will join with an O_2 molecule to form a radical peroxide. A peroxide is two oxygen atoms joined by a single bond, and this peroxide is a radical because when one of the two bonds between the two oxygen molecules was broken in order to form a carbon-oxygen bond, it left an oxygen with only one bond, instead of oxygen's preferred two. The radical peroxide, in turn, can steal a hydrogen from another unsaturated vegetable oil molecule (Fig. 2.17).

These reactions have the nature of a looped chain reaction: A hydrogen breaks off an unsaturated vegetable oil molecule, leaving a radical carbon. The radical carbon combines with an O_2 molecule, and forms a radical, which steals a hydrogen from another unsaturated vegetable oil molecule, leaving another radical carbon, which then combines with an O_2 to form a peroxide, and so on and so on.

Peroxides, two oxygens joined only by a single bond, are a unstable, and will eventually split into two radicals, that, in turn, can both steal hydrogen atoms from other vegetable oil molecules, beginning the process over again. So, because of this peroxide instability, the oxidation of a single vegetable oil molecules can produce three radicals that will and can steal hydrogens from other vegetable oil molecules (the orginal peroxide radical, and two radicals formed by the breakdown of the peroxide). The rates of reactions like these tend to have periods of exponential growth, where the rate is relatively low until some threshold is reached and then the rate of the reaction becomes very fast until the ingredients for the reaction begin to be used up, and the rate of the reaction slows down again.

Peroxide Decomposition and Polymerization. Oxidation is a difficult enough process to understand, but what happens next is almost infinitely complex. The peroxides break apart, leading to a number of different processes that can only be described as a soupy mess of different products that include alcohols, acids, aldehydes, and polymers. To make it worse, temperature, catalysts, and fatty acids all interact in complicated ways to promote different processes and final products. A description of the mechanisms of the formation of the multitiudes of possible products goes well beyond my ability or understanding. [57]

Initiation
Reaction 1: A Hydrogen breaks off an unsaturated vegetable oil molecule, leaving a radicalized Carbon

Energy, Catalyst

Propagation
Reaction 2: An O_2 molecule bonds with the radical carbon, forming a radical peroxide.

Reaction 3: The peroxide radical steals a Hydrogen from another unsaturated vegetable oil molecule, resulting in another radical Carbon and a possible return to Reaction 2

In the diagrams above, dots by atoms represent unpaired electrons.
The curved arrows represent the movement of single electrons in the reactions.

Forest Gregg

Above are two ways of drawing the same fragment. Except for the atoms we care about, the drawing on the right shows only the Carbon skeleton, where every kink is a Carbon. In both drawings, **R** stands for the rest of the stands for the rest of the molecule this fragment is part of.

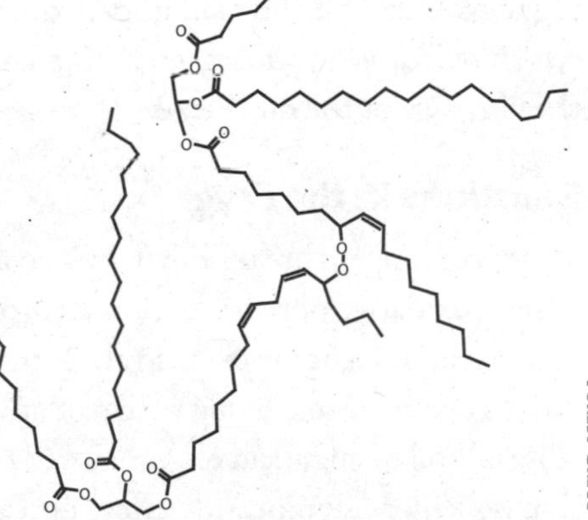

Forest Gregg

Fig. 2.17: *A hydrogen breaks off an unsaturated vegetable oil molecule, leaving a radical carbon. The radical carbon combines with an O_2 molecule, and forms a radical, which steals a hydrogen from another unsaturated vegetable oil molecule, leaving another radical carbon, which then combines with an O_2 to form a peroxide, and so on.*

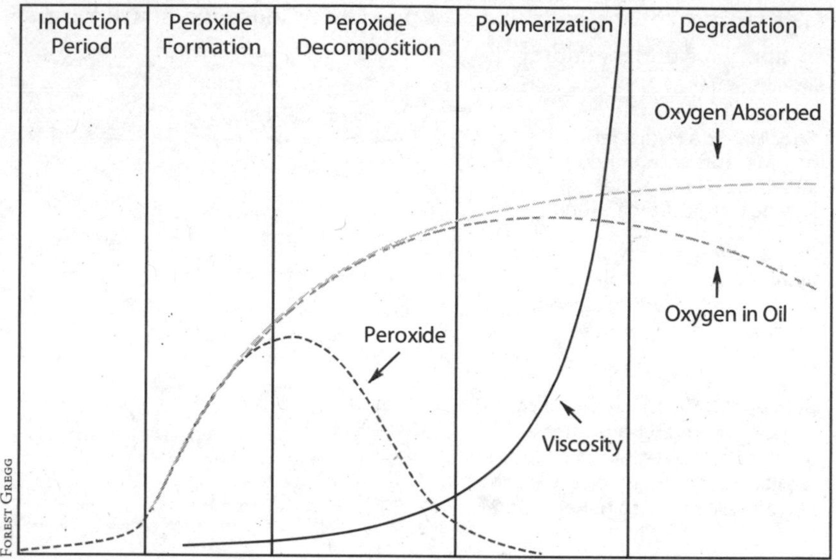

Fig. 2.18: *A fall in peroxides coincides with a sharp rise in oxidative products, including large polymers that ultimately raise the viscosity of the oil.*

Fortunately, the larger scale effects are fairly consistent and clear. The amount of peroxides will increase until it reaches some peak, and then fall, as more peroxides are decomposed than can be formed from the remaining unoxidized oil. The fall in peroxides generally coincides with a sharp rise in oxidative products, including large polymers that ultimately raise the viscosity of the oil (Fig. 2.18).

Reactions in the Fryer

Waste cooking oil breaks down by two additional important mechanisms besides oxidative polymerization: hydrolysis and thermal polymerization. These reactions occur extensively at the temperature and under the conditions of a fryer, and along with oxidative polymerization, are what causes vegetable oil to go rancid once removed from cooking service. Neither reaction is likely to significantly affect unused cooking oil.

As far as fuel properties, these reactions are not as problematic as oxidative polymerization, but hydrolysis does produces products that can form soaps and that increase the amount of water that can be suspended in oil, and thermal polymerization increases the viscosity of the oil.

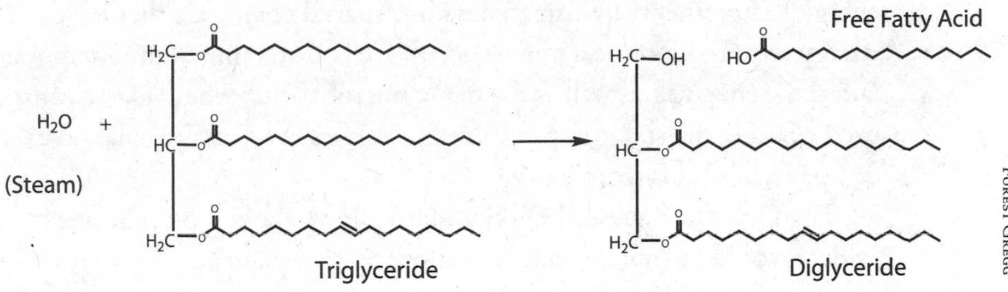

Fig. 2.19: *Hydrolysis is a reaction in which a molecule of steam breaks a fatty acid off of the glycerin backbone of a triglyceride, resulting in a free fatty acid and a diglyceride.*

Hydrolysis

Reaction

Hydrolysis is the name for a reaction in which a molecule of steam breaks a fatty acid off of the glycerin backbone of a triglyceride, resulting in a free fatty acid and a diglyceride (a glycerol molecule with just two fatty acid chains, Fig. 2.19).

Diglycerides can also undergo hydrolysis to form another free fatty acid and a monoglyceride (a single fatty acid attached to a glycerol).

The hydrolysis reaction requires a good deal of energy, and this reaction will usually only take place at or above 212°F, or the boiling point of water.[58]

Acceptable Level of Hydrolysis

Waste cooking oil right out of the fryer usually has acceptable levels of free fatty acids and mono and diglycerides. Even in small concentrations free fatty acids impart an unpleasant smell and taste to oil, so restaurants will stop using oil before it has degraded too much, and the oil will still be acceptable for use as fuel. Typical values of free fatty acids in used vegetable oil seem to be between 1-2 percent.

However, when stored with animal products or in conditions that promote biological growth, hydrolysis can occur extensively through enzymatic means. Animal cells and some bacteria, fungi, and yeast can use triglycerides (vegetable oil), as an energy source. As part of digestion of triglycerides, the fatty acids must be removed from the glycerol, normally a reaction that requires a great deal of energy. Biological cells lower the energy requirements by using enzymes, special chemicals that lower the energy required for a reaction to occur. Cells from animal products contain these enzymes, as well as the many microbes that would like to eat the used cooking oil, so improper storage can cause very advanced, and even complete, levels of hydrolysis.

The German Rapeseed Oil Standard allows about 1 percent free fatty acids. I would be more liberal, and allow up to 3 percent.

Problems

Attempts to use oil with free fatty acids concentrations over 15 percent have been largely unsuccessful, with reports of frequently clogged filters, and buildup of gummy deposits in injection pumps and injectors. Presently, we don't have a very good understanding of why these high free fatty acid oils caused these problems. It's possible that it doesn't have anything to do with the products of hydrolysis, since oil that has been so deteriorated through hydrolysis has almost certainly also degraded through other mechanisms.

At lower concentrations, we have a better understanding of how free fatty acids and mono/diglycerides can affect fuel properties. Free fatty acids are mainly of concern since they can form soaps in the presence of bases and metals. Some soaps are excellent in binding oil and water together, a property that is very useful for washing oily pots and pans, but less attractive for fuel. Other soaps have nearly opposite properties, forming deposits that can't be dissolved in water or oil, for instance soap scum in a bathtub. Soapmaking is a potential way to lower free fatty acids in fuel preparation, by adding calcium in the form of quicklime or slaked lime to form insoluble soaps that will precipitate out of solution and be more easily removed.

Fatty acids are also minor prooxidants. That means they slightly increase the rate of oxidative polymerization.

Mono and diglycerides, the other products of hydrolysis, are also extremely effective in binding water into vegetable oil fuel. Indeed they are produced industrially for just that purpose and are a common ingredient in processed foods and cosmetic creams, and used extensively in the oil industry. Monoglycerides and diglycerides are not very reactive, and there is no practical way that I know of to separate them out of vegetable oil.

Determining Level of Hydrolysis

Borrowing from the biodiesel field, free fatty acids can be determined by something called titration. I won't go into details here, because the information is easily found in any beginning biodiesel book[59] and is excellently described many places on the Internet, but basically it depends upon a measure of how acidic the oil and the result is called the Acid Number or Acid Value. The percentage of free fatty acids will be close to *half* the Acid Number.

For a more subjective test, high free fatty acid oils usually stink to high heaven. While pulling oil out of a dumpster should be avoided for a number of reasons, if the smell of the oil makes your stomach upset, it's probably high in free fatty acids along with many other nasty things.

Recommendations

Problems with hydrolysis can usually be avoided just by arranging to pick up the oil soon after a restaurant removes it from the fryer or by arranging for the oil to be stored where it won't be exposed to air, water, and light. If you have suspect sources of oil, a titration is in order. Personally, I would avoid using oil that had a concentration greater than 3 percent free fatty acids.

Yellow and brown grease are the products produced by the rendering companies that own most of the oil dumpsters behind restaurants. Renderers pick up the waste oil from those dumpsters and assorted food wastes from food processors and mix it all together, clean it up somewhat, and sell it on the commodity market. With increased demand for biofuels, prices for yellow grease are going up, but you used to be able to buy about 5,000 gallons of the stuff for 14 cents a pound (about a dollar a gallon). Yellow grease is not a good fuel for vegetable oil conversions for a number of reasons, but one important one is that typically they have very high free fatty acid values, as high as 15 percent.

As far as reducing free fatty acids and mono/diglycerides. There are not very many known good solutions, but we will discuss what has worked and what might work in the section on Dewatering and Filtering in Chapter 4.

Thermal Polymerization

We've already talked about oxidative polymerization, and how it is dramatically accelerated by heat. High heat, above 300°F, can also cause vegetable oil to polymerize, through a different mechanism that doesn't require oxygen.[60]

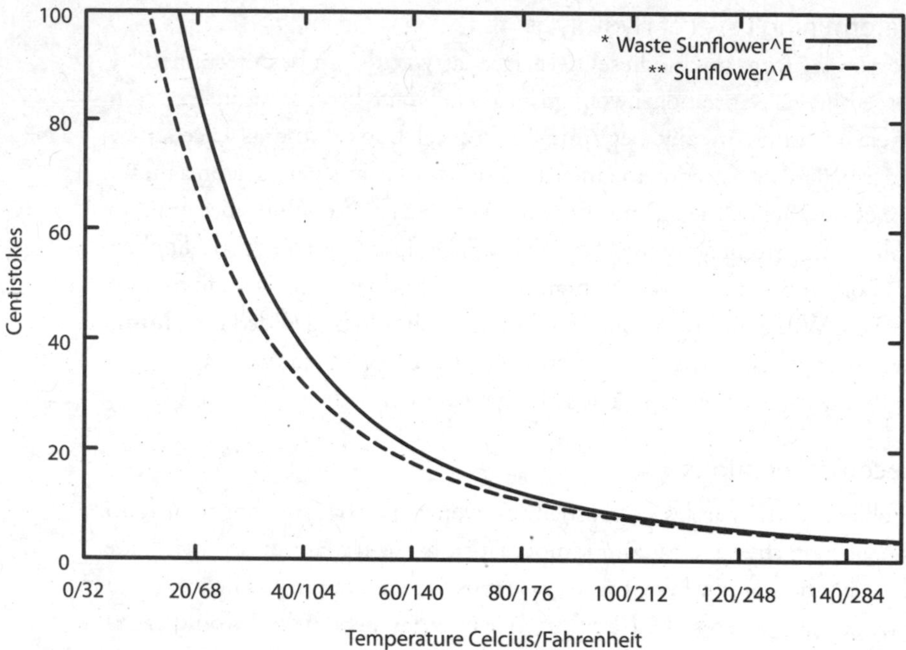

*Pugazhvadivu M., and K. Jeyachandran, "Investigations on the performance and exhaust emissions of a diesel engine using preheated waste frying oil as fuel," *Renewable Energy* 30, no. 14 (2005): 2193, dx.doi.org/10.1016/j.renene.2005.02.001
**Ryan T.W., L.G. Dodge, and T.J. Callahan, "Effects of vegetable oil properties on injection and combustion in two different diesel engines," Journal of the American Oil Chemists' Society 61, no. 10 (1983): 1615

Fig. 2.20: *Viscosities of used and unused sunflower oil. High heat causes vegetable oil to polymerize and thicken once cooled.*

This mechanism is called thermal polymerization. At frying temperatures, vegetable oil undergoes complicated and poorly understood reactions that result in vegetable oil molecules splitting apart and clumping together as polymers. The bits that break apart are lighter and tend to evaporate from the oil. These fragments make up a big part of the smell of frying. What's left when the oil cools down are the larger molecules — the unaffected vegetable oil molecules and the polymers — which makes used cooking oil thicker than unused cooking oil (Fig. 2.20). This is not of great concern for conversion systems that adequately heat the oil, because the viscosity of used cooking oil will converge with the viscosity of fresh oil, if both are heated enough.

Recommendations

Thermal polymerization stops when the oil is cooled, and is not that great a problem for fuel properties.

Water

Water is a common contaminant of both petrodiesel and vegetable oil fuel that allows microbial growth in the fuel, promotes the chemical corrosion of metal components of the fuel system, and can cause accelerated wear in high pressure pumps and injectors. Total water content should be no more than .07 percent by volume. Since exact water content is difficult to determine, vegetable oil should at least pass a crude field test called the crackle test. We explain how to remove water in the section on Filtering and Dewatering.

Water and Oil

Water can be present in oil in four forms: free, suspended, emulsified, and dissolved.

Free Water

When mixed together, water and oil will tend to separate from one another and form distinct layers. You've probably noticed this with in oil and vinegar salad dressings that you have to shake before using, and which will separate again in the refrigerator (vinegar is basically just acidic water). We will use the term free water for water that has separated into a distinct layer underneath the oil or for water that will settle into such a layer within minutes. Free water causes the most problems but it is also the easiest to remove.

Suspended Water

As an oil and water mixture is agitated, the water will form smaller and smaller droplets. Once the agitation stops, these droplets can take a long time finding one another, coalescing, and separating out as free water. Water that is mixed into oil as very fine droplets and that are slow to separate out, is called suspended water.

Emulsified Water

If fine droplets of suspended water are coated with molecules called emulsifiers, then the water droplets will no longer be attracted to each other, and may never coalesce and form a separate layer. Water in this form is called emulsified water, and this stabilized mixture of oil and water is called an emulsion. Mayonnaise is the classic example of an emulsion, being a mixture of oil and vinegar stabilized by emulsifiers found in egg yolk. It can be quite difficult, if not impossible to remove all emulsified water from vegetable oil. On the other hand it is not clear how much danger tightly emulsified water poses to diesel fuel systems.

Dissolved Water

Dissolved water is something like a Big Foot in the vegetable oil community. Passion about the subject seems to be directly related to how little evidence there is that it exists. Well, I believe it exists, but just barely. At room temperature, water is soluble in vegetable oil at the level of 750 micrograms per liter, which is .00008 percent by volume, or 0.8 parts per million.[61] That's not very much. As far as danger to the fuel system, I would be more worried about the potential damage caused by Big Foot darting out in front of your car on a dark country lane.

Water Damage

Biological Activity

Free water that has settled to the bottom of a fuel tank, fuel filter, lift pumps, and injection pump can provide a home for a class of fungi and bacteria called hydrocarbon utilizing microorganisms, also known in the fuel industry as HUM bugs. Often mistaken for algae, HUM bugs feed off hydrocarbons like diesel fuel or vegetable oil. HUM bugs are a problem because they form dark, gel-like mats that can clog filters, consume seals and gaskets, and excrete acids that can damage metal components of the fuel system. The telltale sign of infestation are filters clogged with what looks like black, brown-green or reddish sludge, algae, or coffee grounds. Fortunately, HUM bugs tend to only be a problem in fuel systems that are left stagnant for extended periods of time, and so the best way to avoid problems is to just not leave your vehicle sitting for weeks on end. If you do have an infestation, you should clean your tank and treat your fuel with biocides available at your local auto parts store. We've had good results with Soltron[62], but any biocide designed for diesel fuel should work.

Oxidation

Free water will promote the oxidation and corrosion of metal components of a fuel system, especially since the water is likely to be at least somewhat acidic. Acids may have been introduced into the oil while it was used for cooking, or free water may combine with the sulfur from diesel fuel to form sulfuric acid.

Loss of Lubricity

The pumps and injectors of a diesel fuel system rely on the fuel to lubricate these components' finely mated surfaces. Water is a lousy lubricant, and tiny droplets of suspended or emulsified water can disrupt the fuel's lubricating ability, allowing metal surfaces to rub, scuff, and scratch against each other. As the clearances between mated surfaces are degraded, the performance of the component will worsen and ultimately fail.

Cavitation

The boiling point of water depends upon pressure. This is how a pressure cooker can heat water well over 212°F without boiling by increasing the pressure of the vessel, and how at high altitudes water will boil at a temperature below 212°F due to lower atmospheric pressure. In a diesel engine, the action of high pressure pumps and fuel injectors produce intense waves of pressure that pass through the fuel, and any water suspended in that fuel. In the trough of one of these pressure waves, pressure might be low enough so that a droplet of water will turn from a liquid into a bubble of gas that is much larger in volume. Once the pressure begins to climb again, the gas bubble will very quickly condense back into

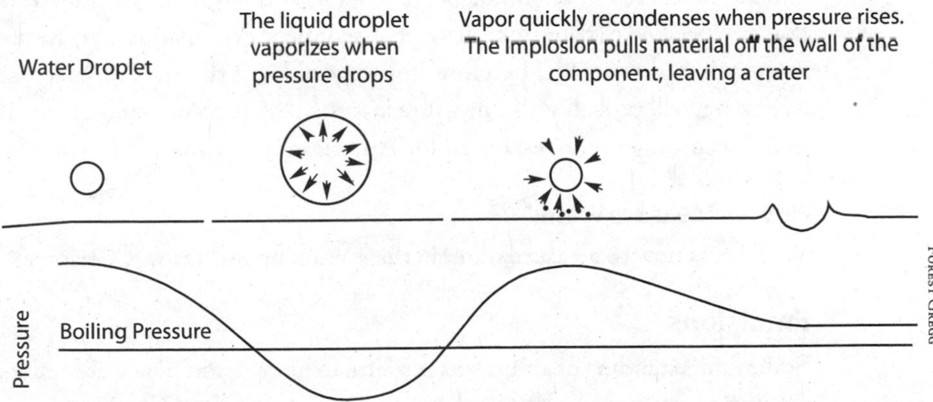

Fig. 2.21: *Cavitation, a type of damage caused by water, is a concern with vegetable oil fuel.*

Forest Gregg

a small liquid droplet. That change from gas to liquid, and the corresponding change in volume, can happen so quickly that it creates enough force to blast off bits of surface from the pump or injector. This kind of damage is called cavitation (Fig. 2.21).

Of all the possible types of damage caused by water, cavitation seems to be the one that most concerns the vegetable oil community. I tend to think that the danger is a bit out of proportion.

How Much Water?

We follow the limits of water that are laid out in the German Rapeseed Oil Standard of .075 percent water by mass, which is about .07 percent by volume. Vegetable oil fuel has gained a greater measure of mainstream acceptance in Germany than any other country, and this standard, in its present form, was developed to control the quality of commercially sold vegetable oil fuel by a consortium that included a number of universities, the Bavarian Ministry of Agriculture, diesel engine manufacturers, agricultural equipment makers, and vegetable oil producers and traders. In the seven years that the standard has been used, the lead developer claims that water content at the specified level has never been a problem.

American standards for diesel and biodiesel only allow .05 percent water and sediment, so it may appear that we are calling for a significantly higher allowable content of water in vegetable oil than what is acceptable in American diesel and biodiesel, but that is not the case. In the US, diesel and biodiesel are limited to .05 percent water and sediment, but the tests that are specified to determine water and sediment do not accurately measure emulsified water.[63] Therefore, any vegetable oil that has .07 percent *total* water, will probably also measure less than .05 percent water and sediment according to the tests used for American diesel and biodiesel.

How to Measure Water

We discuss how to measure water in the section on fuel testing, Chapter 4.

Emulsions

Some understanding of emulsions is useful to understand how water exists in oil and how it can be removed. So, now we are going to talk about mayonnaise because (1) it's my favorite condiment, and one I hope you are familiar with, and (2) it is the classic emulsion, in my fawning opinion. If we can understand mayonnaise, we can understand water in oil.

Let's start by looking at why oil and vinegar don't like to mix. First off, vinegar is just acidic water, and for our purposes, we can just treat it as water. Water is a polar molecule. That means that a water molecule always has a region that is positively charged and other regions that are negatively charged. The regions of positive or negative charge are called poles (see Fig. 2.22). Just like oppositely charged poles of two magnets are attracted to each other, so are the oppositely charged poles of two water molecules. This attraction is called *polar attraction*, and it is a pretty strong kind of intermolecular attraction.

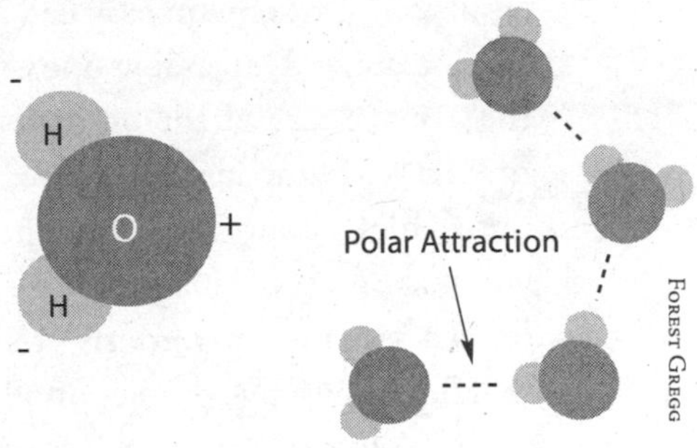

Fig. 2.22: *Water is a polar molecule; vinegar is essentially acetic water.*

Vegetable oil is almost entirely nonpolar; it has no regions of strong permanent charge. This does not mean that vegetable oil molecules are not attracted to other molecules. They are, but with a much weaker force called London force. London forces are interesting and complicated, but all you need to know right now is that they produce a much weaker attraction between molecules than a polar attraction (see the section on viscosity in this chapter for more details).

The reason that water and oil don't like to mix is not that they aren't attracted to each other; it's that, because of polar attraction, water is much more attracted to water than it is to vegetable oil. Given an opportunity, water droplets will clump together and sink below the vegetable oil, and vegetable oil droplets will clump together and rise above the water (water will sink below oil because it's denser).

Mayonnaise works because it doesn't give oil droplets a chance to get together. (Mayonnaise is actually an oil in water emulsion, but the general principles hold for water in oil emulsions, such as we are concerned about in vegetable oil fuel.) The first step to create an emulsion is to mix the heck out of the water oil mixture so that the oil is reduced to very small droplets dispersed throughout the vinegar. This is called a mechanical emulsion, also sometimes called a suspension, and it is a successful short-term strategy. Depending upon how much oil there is in the vinegar and the average size of the oil droplets, it can take a long time for the oil droplets to bump into each other, cohere, and get large enough to separate out. Think about an oil and vinegar salad dressing. The more vigorously you shake it up, the longer it takes for the oil and vinegar to separate. That's because the more vigorously you shake the dressing, the smaller you are making the average droplet of oil.

The second step and the secret to mayonnaise is the use of an emulsifier, specifically lecithin from egg yolk. An emulsifier is a molecule that has parts that are polar and parts that are nonpolar (Fig. 2.23). In our oil and vinegar mixture, molecules of emulsifier have their nonpolar portion inside the droplet of oil and their polar end in the vinegar. If there are enough emulsifiers, they will create a polar shield around the nonpolar droplet of oil that prevents another droplet of oil from getting to the emulsified droplet and cohering.

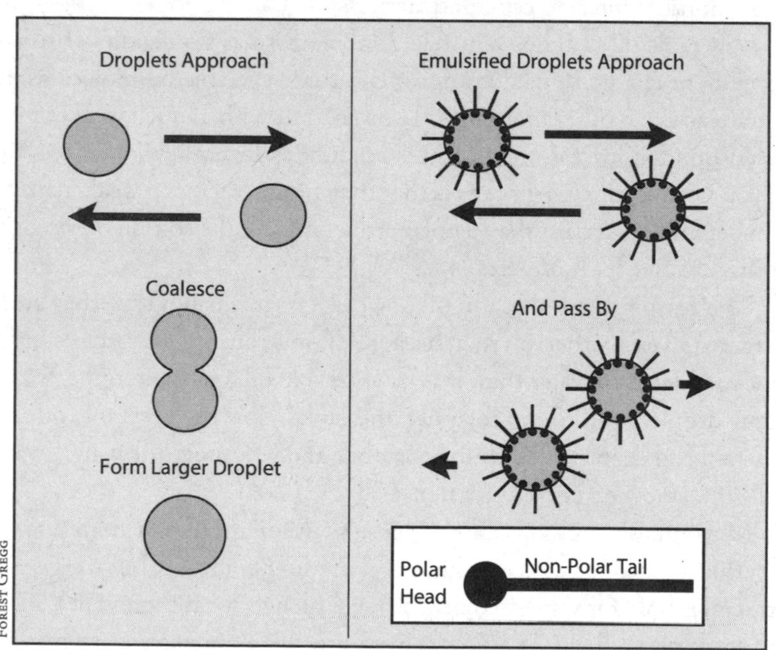

Fig. 2.23: *An emulsifier is a molecule that has parts that are polar and parts that are nonpolar.*

Emulsifier-stabilized emulsions are called chemical emulsions, and it can be very difficult to break such an emulsion and get the oil to separate from the water. Typical strategies are to heat or cool the emulsion or to introduce a dispersant, a chemical that neutralizes the action of emulsifiers.

Turning to emulsions of water in used cooking oil, the principles are the same. The stability of the emulsion depends upon the average size of the water droplets, the total amount of water in the oil, and the presence of emulsifiers. Unfortunately, as oil is used in frying, excellent emulsifiers are produced as vegetable oil molecules are broken down into molecules called diglycerides and monoglycerides though hydrolysis (see high-temperature reactions). Diglycerides and monoglycerides are such powerful emulsifiers that they are actually produced on an industrial scale for use commercial food industry.

As of this writing, heating is the only technique used to break emulsions of water in used cooking oil commonly used in the vegetable oil fuel community. Most plans for dewatering oil have some provision for moderately heating the oil. While it has been proposed a number of times, there is no widespread use of chemical dispersants, although this is certainly an avenue that calls out for investigation.

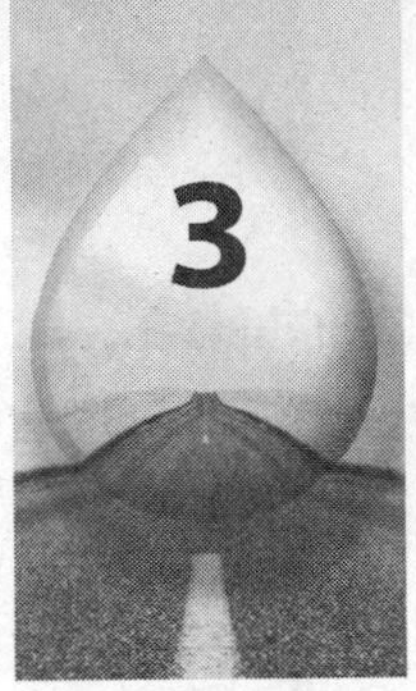

System Design

At the most basic level, a vegetable oil conversion is just a heated fuel system. By fuel system, I mean a fuel tank, fuel lines, filters, and pumps — everything before the injection pump or high-pressure pump. Some conversions, called one-tank systems, alter the stock fuel system to add heat. We advocate the two-tank approach, which means leaving the stock fuel system as unaltered as possible and installing an entire secondary fuel system designed from the ground up for vegetable oil. Besides heating the fuel, two-tank conversions must also control what fuel is feeding the engine, and where the returning fuel is being sent.

In order to fulfill these functions, most two-tank systems have these components.

1. Heated fuel tank
2. Heated fuel lines
3. Heated vegetable oil filter
4. Final fuel heater
5. Valves to control what fuel is being burned
6. Manual or automatic control of valves
7. Indicators for the driver of the status of the system.

Depending upon the vehicle and application, the conversion system may also include an auxiliary pump dedicated to vegetable oil fuel.

Right now, there are a lot of differences between how different companies and individuals build these components, and even if all these components are included in the conversion systems. Ultimately, these differences are a reflection of two related things: different design priorities and different levels of skill and knowledge on the part of the designers. Design priorities include cost, longevity, reliability, ease of installation, redundancy, safety, and more. Every design involves trade-offs between these different priorities. As an old friend used to say, "You can have it good and cheap, cheap and fast, or fast and good. But you can't have all three."

The skill and knowledge of the designer is important because a good designer knows the price of things. He or she understands where and what the trade-offs are, and sometimes the designer can identify ways to reduce costs of particular trade-off. With a very skilled designer, sometimes you can get good, cheap and not too slow; cheap, fast and not that bad; or fast, good, and not that expensive. With a bad design, you will get bad, expensive, and slow.

Putting the question of the abilities of the designer aside, it's hard to say that one set of design priorities is better than another, but it is important if you are selling systems to customers that you inform them what your priorities are, so they can see if they match their own.

There are a number of companies that sell very cheap systems that are unreliable, and cause serious damage to a high percentage of engines within 20,000 miles. There are also customers that bought a junker for $700 and if they can run on "free" fuel for 10,000 miles they will be very happy. It doesn't make sense to these customers to buy a more expensive system, since they expect the car to blow up at any moment anyway.

The problem comes in when a company represents their product as being other than what it was designed to be to potential customers who don't have the knowledge to allow them to evaluate these claims.

In this section we are going to discuss how we think a conversion system should be designed, based upon what we've discussed about diesel engines and the fuel properties of vegetable oil. We can now investigate how a vegetable oil conversion system should be designed for our personal design priorities. Our priorities may not be the same as yours, but hopefully after this discussion you will be able to evaluate a kit to see if it meets your priorities, or make choices that better fit your priorities if you are building your own system.

Our priorities:

1. Do no harm.
   a.) The conversion should not shorten the useful life of the engine or vehicle.
   b.) The conversion should not prevent the operation of the vehicle on the stock petrodiesel fuel.
   c.) The conversion should not complicate regular maintenance to the engine.
   d.) Safety should not be compromised.
2. The vehicle should run as well as technically possible on the alternative fuel.
3. The system should last the life of the vehicle.
   a.) Moving and wearing parts should be used as little as possible.
4. Likely points of failure should fail gracefully and obviously.
5. Common user errors should be identified, and where possible, designed around.
6. An average car owner should have no difficulty operating a vehicle that has been converted.
7. An average car owner must be able to convert the vehicle themselves.
8. An average car owner must be able to afford the components of the conversion themselves.

Now, here's where knowledge comes in. We can apply what we've discussed about the workings of diesel engines and the fuel properties vegetable oil, and from our experience with conversions come up with a set of design principles we must meet to achieve our above stated goals. Here we go:

1. Vegetable oil fuel must be injected at temperature of 160°F or higher (see the section on Viscosity).
2. Vegetable oil fuel should not be injected until the combustion chamber has reached operating temperature (see the sections on Wet Stacking, Combustion, and Oxidative Polymerization).
3. Vegetable oil fuel should not contaminate the lubricating oil (see the section on Oxidative Polymerization).

4. Vegetable oil fuel should not contaminate the diesel fuel supply (see the sections on Viscosity and Cold Flow Properties).
5. Materials that are incompatible or reactive with vegetable oil should not be used (see the section on Oxidative Polymerization and the section on Solvency in Other Fuel Properties).
6. The fuel system should be heated to allow for the smooth flow of vegetable oil fuel, but unnecessary heat should be avoided so as to not accelerate oxidation (see the section on Viscosity and on Oxidative Polymerization).
7. The size of fuel lines and fuel filters should be as large as practical to decrease the pressure or vacuum necessary to move vegetable oil fuel through the fuel system (see the section on Viscosity).
8. Auxiliary pumps must be carefully chosen for their ability to pump vegetable oil (see the section on Viscosity).
9. Electrical components should be used with an eye to the capacity of the vehicle's alternator and the associated load on the engine (see the section on Electrical System).

These are the principles that we used to design the Frybrid system, and at the end of this section we will have described a system that looks a lot like the product our company offers. However, although the Frybrid system is based upon these principles, these principles are not based upon the Frybrid system; i.e., the principles were not designed to prove that only our products were acceptable. There are a few other companies that also understand and apply these principles and sell quality kits, and there are a great many people who have used these principles to design homebrew systems that work very well. Unfortunately, most of the kits that are sold today fail to apply even the most critical principles.

Now, we turn to the most basic question — One Tank or Two?

One Tank or Two?

Probably the most basic difference between conversions systems is whether they alter the vehicle to be fueled solely by vegetable oil or whether the system uses both stock fuel and vegetable oil. This difference is usually referred to as one-tank or two-tank systems, respectively. In a one-tank system, the vehicle starts on vegetable oil, runs on vegetable oil, and shuts down on vegetable oil. In a two-tank system, the car starts on petrodiesel or biodiesel, switches to vegetable oil, and then shuts down on petrodiesel or biodiesel. We cannot recommend using a one-tank system unless your engine was specifically built to burn vegetable oil fuel.

Application of Principles

One-tank systems violate our first three principles: (1) Vegetable oil fuel must be injected at temperature of 160°F or higher, (2) Vegetable oil fuel should not be injected until the combustion chamber has reached operating temperature, and (3) Vegetable oil fuel should not contaminate the lubricating oil.

At startup it is difficult to ensure that the vegetable oil fuel is adequately heated, and the high viscosity of the cold oil will lead to a poor spray pattern that, in the unheated combustion chamber, will lead to incomplete combustion and coking as well bypass of liquid fuel past the unsealed rings, and into the engine lubricating oil.

Even if it were possible to adequately heat the oil prior to start up, significant amounts of fuel will contaminate the engine oil until the engine reaches operating temperate and the rings seal.

With a two-tank system, the engine starts on the stock fuel and should run on the stock fuel until both the engine is at operating temperature and the vegetable oil can be heated to injection temperature. In a maintained engine, the stock fuel will burn with an acceptable level of completeness at startup. Although the petrodiesel will bypass the rings and contaminate the lubricating oil until the engine warms up and the rings seal, modern engine oils are formulated to handle typical levels of contamination by petrodiesel.

Engines Designed to Burn Vegetable Oil

You may have heard that Rudolph Diesel designed his original engine to run on peanut oil. While that is not true, it is certainly possible to build an engine designed specifically for the fuel properties of vegetable oil, and a few companies have done so through the years. These engines are difficult to find, but they do exist. We only recommend one-tank systems with these types of engines.

Engine Design and Modification

Because the long-term performance of an engine depends upon how the fuel and engine interact, it is not surprising that different designs of conventional diesel engines vary in how accepting they are of vegetable oil fuel, and that it is possible to design, from the ground up, an engine able to burn unheated vegetable oil, or modify a conventional engine to make it more accepting of the alternative fuel.

Indirect and Direct Injection Engines

In general, indirect injection engines are more tolerant of inadequately heated vegetable oil than direct injected engines, for a number of reasons.

The first, and probably most important, is that the proper mixing of air and fuel in an indirect injection engine is not nearly as dependent on the spray pattern as it is in direct injection engines. In indirect injection engines, most of the mixing is accomplished by the partial combustion of fuel forcing the air/fuel mixture into the main combustion chamber. In a direct injection engine, the quality of the spray pattern is the most important factor in efficiently mixing fuel and air, so these engines are much more sensitive to the higher viscosity of inadequately heated vegetable oil and the poor spray pattern that viscosity causes.

Second, cylinder temperatures are typically hotter in an indirect injection engine because of higher compression ratios. The higher temperatures mean that vegetable oil will be exposed to a longer period of temperatures sufficient for combustion.

Third, indirect injection systems tend to use pintle-style injectors which have a self-cleaning action that makes them less susceptible to problems caused by carbon buildup than the hole-style injectors typical of direct-injection engines.

Designing an Engine to Burn Vegetable Oil

As mentioned before, the problems associated with using vegetable oil in a conventional diesel engines are not the "fault" of the vegetable oil, but the result of a mismatch between the fuel and the design of the engine. In the next chapter we will be discussing how to alter the fuel to bring it into alignment with a conventional diesel engine, but before we do, we will look over some attempts to align the engine to the fuel. ☞

The Elsbett Engine

The most prominent instance of an engine designed particularly for vegetable oil was the Elsbett Engine developed by Ludwig Elsbett and demonstrated in 1973. The engine had a combustion chamber design that reduced the heat transfer to the cylinder walls and piston, which allowed the cylinder to be adequately cooled through oil channels instead of the conventional glycol/water cooling system. This allowed the average temperature of the piston and cylinder walls to be much hotter than in an engine with a conventional cooling system, which led to many benefits in attempting to burn thicker, slower to evaporate, slower-burning fuels such as vegetable oil. The most significant benefit of the hotter temperatures was keeping the surfaces in the cylinder too hot for carbon deposits to form upon them.

Elsbett Combustion Chamber Design

The key to the Elsbett Engine was a combustion chamber design that was composed of:

1. A deep spherical combustion chamber in the top of the piston.
2. Valves and piston designed to form a rapid swirl of air in the spherical combustion chamber.
3. Injectors aligned to spray the fuel in line with the direction of the air swirl.

The swirl of air in the spherical combustion chamber produced a centrifugal effect that resulted in cooler denser air being pushed towards the wall of the combustion chamber, while the hottest air was concentrated in the middle. The spray pattern did not disrupt this stratification of air temperature. The result was that the cooler air acted as insulator to the piston and cylinder walls, radically reducing the heat transferred to these components.

Engine Modifications

Conventional engines can be modified to be more accepting of vegetable oil. In mechanically controlled systems, timing injectors can be modified, and injection pump timing can be adjusted. In electronically controlled systems, it may be possible to control every aspect of injection through changing the electronic fuel maps. In both types of systems, the goal of the modifications should be to ensure proper closing ☞

of the injectors, advance the end of injection, and, if possible, increase the injection pressure.

Mechanically Controlled Systems

The easiest adjustment for most mechanical systems is to adjust the beginning and end of injection. Advancing the beginning of injection will bring some benefit in efficiency of combustion on either fuel, but the big pay-off for long-term performance and durability is advancing the end of injection. It takes longer for vegetable oil to burn than diesel fuel, which means that if the end of injection is optimized for diesel fuel, then more vegetable oil is being supplied to the cylinder than can burn completely. The "smoke screw" is the usual name for the adjustment that controls end of injection.

Swapping injectors can bring some small benefits in long-term performance. "Economizer" injectors — injectors that have higher opening and closing pressures — can help with secondary injection and nozzle dribble as they have stronger closing spring pressure.

Electronically Controlled Systems

Electronically controlled systems are much more difficult for the average person to adjust, but the fine control of timing and pressure by electronic control represents a huge potential for aftermarket research and development. Elsbett already includes reprogrammers as part of some kits, and this avenue will become more important in the coming years. ■

Ludwig Elsbett designed an engine specifically to burn vegetable oil fuel, known as the Elsbett or Elko engine. Over 1,000 engines were manufactured from 1979 until 1994, when Elsbett Konstruktion, the manufacturer of the Elsbett engine, was sold after financial difficulties. The engines sometimes come up for sale, usually with a vehicle wrapped around it. AMS Antriebs-und Maschinentechnik of Schönebeck, Germany licensed Elsbett engine technology and produced an engine designed for stationary applications. MWB Motorenwerke Bremerhaven AG, also of Germany, produces an engine designed to burn vegetable oil, based upon a design effort led by Dr. Kampmann and funded by the German government. W. Mahler AG of Switzerland developed a very interesting design for a multifuel engine capable of burning vegetable oil, but has not produced it for sale as of this date.

Elsbett Konstruktion licensed its technology to companies in Malaysia and the former Soviet Union, though I have not been able to track those companies down.

One-Tank Conversions

One-tank conversion systems are very popular, especially in Europe. However, there is no one-tank conversion system that adequately addresses the problems of cold starting on vegetable oil, and that includes the system currently being marketed that includes engine upgrades such as new injectors, different glow plugs, changes in timing, and computer reprogramming.

Based upon the research literature and our experience, we expect that on average, two-tank systems should lead to fewer problems over time.

Climate

One-tank systems are completely inappropriate for climates where the temperature regularly falls below 40°F, or when using thicker oils regardless of climate, since the oil left in the high-pressure fuel system will be too thick to be adequately sprayed into the combustion chamber and ignite, and the vehicle will not start. Blending petrodiesel into the fuel or running a heating system overnight from an mains power can address this problem, but doing so brings much higher operating costs.

Heating

In a vegetable oil conversion, heating has two functions. The first is to heat the oil enough so that it will flow through the fuel lines and filter. The second is to reduce the viscosity of the fuel before it's injected in order to achieve good combustion. This first function is important, but the second is the single most important task of vegetable oil conversion system. Heating vegetable oil to achieve an acceptable viscosity at injection is the heart of a successful conversion.

While heat is necessary, higher temperatures also accelerate the oxidative breakdown of vegetable oil, so care should be taken not to overheat the oil in the tank.

Heating Methods

There are two main sources of heat that can be used to raise the temperature of vegetable oil: the waste heat of the engine captured in the engine coolant and electricity. Exhaust heat has been proposed many times as a heat source, but so far no one has been able to overcome the problems of extremely high temperatures, great variability in the temperature, and problems of exhaust condensation.

Application of Principles

There are a number of principles at play here:

1. Vegetable oil fuel must be injected at a temperature of 160°F or higher.
2. The fuel system should be heated to allow for the smooth flow of vegetable oil fuel, but unnecessary heat should be avoided so as to not accelerate oxidation.
3. Electrical components should be used with an eye to the capacity of the vehicle's alternator and the associated load on the engine.

First, we will describe what we think works well, and then we will discuss some other approaches.

Recommendations

Our system relies mainly on coolant heat for the following reasons: (1) a typical diesel engine already regulates the temperature of coolant heat to about 180°F, a temperature that makes it easy to reach the minimum injection temperature through heat exchange, (2) heat exchangers do not depend upon any moving parts, will not burn out, and will not short out, so there is no risk of burning the oil or starting a fire, and (3) using coolant adds nearly no additional load to the engine.

We recommend using engine coolant to moderately heat the fuel tank, fuel lines, and fuel filters to ensure that oil can flow fairly easily through the fuel system, and using a very efficient final fuel heat exchanger to boost the temperature of the oil to injection temperature, 160°F or better. We also use a controlled loop to prevent heated vegetable oil from returning to the tank and overheating the oil, where it is at the greatest danger of oxidative breakdown. Electric heat has a place in heating components difficult to heat though coolant means, and in boosting the temperature of vegetable oil above the temperature of the coolant.

Final Fuel Heat Exchangers

It is our experience that an efficient final engine coolant/vegetable oil heat exchanger is critical to bring the oil to the minimum injection temperature. We have used two types of final fuel heat exchangers, flat plate and coaxial. The two types have about the same efficiency in heat exchange, but the flat plates are more compact and more expensive, while the coaxial types are larger and cheaper. Which one we use depends upon the application. Shell and tube heat exchangers can work as well, but they tend to need to be much larger in order to match the efficiency of the other types, so we do not use them for this reason.

Heated Fuel Lines

To heat the fuel lines, we use what we call hose-in-hose, an aluminum fuel line running inside a jacket of hot coolant (Fig. 3.1). We use an aircraft-rated, 3000 series bendable aluminum tubing for our fuel line, and have never

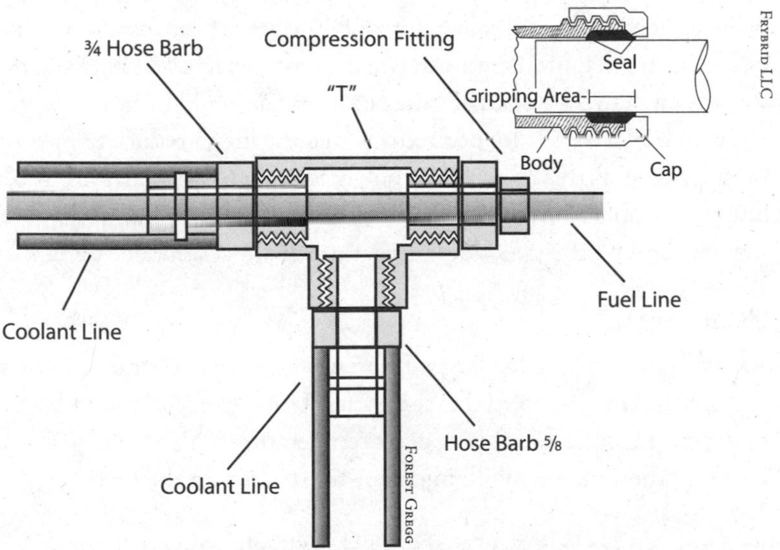

Fig. 3.1: *The hose-in-hose fitting, an aluminum fuel line running inside a jacket of hot coolant.*

had a problem with a leak forming inside the hose-in-hose line, or contamination of the fuel by the coolant. We do not recommend hose-in-hose that use cross-linked polyethylene (PEX) as the fuel line, as there have been problems with PEX line becoming brittle and fracturing. Stainless steel is an acceptable material, but more expensive and difficult to work with than aluminum, as well as more reactive with vegetable oil.

In-Tank Heating

The amount of heat that you need in your tank very much depends upon your climate and what vegetable oil you are using. We built the Frybrid In-Tank Heater so that our system can be used in every climate and with every fuel, and we have a number of customers who have been using our system with great success in Alaska and the Yukon at temperatures below -15°F. It consists of fluted aluminum coolant coils that surround two aluminum draw and return tubes, all of which is surrounded by an aluminum shell that serves to concentrate the heat and act as a fuel reservoir.

In warmer climates, less heat is needed, and in semitropical or tropical areas no heat may be necessary, depending on what oil you are using. Even in hot climates, heat in the tank is probably necessary to use hydrogenated oil, palm oil, or animal fats. Temperatures of 70°F are sufficient for most cooking oils to ensure that fuel is easily drawn into the fuel line, but partially hydrogenated oils and semi-solid oils such as palm oil should be heated to 90°F.

We have moved to looped returns mainly due to reduce temperatures in the tank and risks of oxidative polymerization. By preventing 160°F or hotter vegetable oil from returning to the tank, we have found that even in the hottest days the tank does not get much above ambient temperature.

Electric Heat

We use electric heat for larger filters that are impractical to heat with coolant, and in air-cooled diesel engines. Using electric heat to boost the temperature of the oil above coolant temperature is also attractive, and we plan on experimenting with injection-line heaters in the future.

Another Approach: No Final Fuel Heat Exchanger or Heater

There are some systems that do not have a final fuel heat exchanger or heater, but just heated filters, heated fuel lines, and heated tanks.

Sometimes this is justified by claiming that the injection pump, injection lines, or the engine itself acts as an efficient final heater. For a system such as the late-1990s Powerstroke, where the fuel rail is inside the head, there is actually a good bit of sense to this argument. However, in systems with an injection pump and external injection lines such as the mid-1980s Mercedes, this argument holds no water whatsoever. In these types of systems, without a final fuel heater, the vegetable oil will not reach the adequate injection temperature of 160°F.

Yet Another Approach: Electric Final Fuel Heater

Electrical elements can heat the vegetable oil to an adequate temperature, but at a cost of more moving parts, possible additional sensors and switches to prevent the element from overheating, and a greater load on the alternator.

Compared to coolant heat exchangers, the effectiveness of electric heating elements is much more sensitive to changes in flow rate of the fuel, temperature of the fuel, and air temperature. Electrical elements must be chosen carefully to be able to adequately heat the oil under high flow conditions and not overheat when the flow rate drops.

Positive temperature coefficient (PTC) elements are strongly preferred because of the self-limiting properties of these types of heaters. These types of elements are pre-set at the factory for a maximum temperature they will reach even in no-flow conditions. This makes them much safer to use than simple heating elements like nichrome wire or glow plugs, where maximum temperature is a function of resistance, voltage, and heat shedding ability, and errors in design can lead to catastrophic damage of components of the vehicle and even fire.

Load Cost of Electric Elements

The additional load of electric elements may be more than your charging system can handle leading to dim headlights and dead batteries in the short run, or shortened alternator life in the long run.

Typically, a vehicle's alternator should be rated to produce 1.5 times the expected load of a vehicle. This is another way of saying that in a typical alternator two-thirds of the rated amps are going to the regular load of the vehicle, and one-third is available for recharging the batteries or for intermittent high-draw loads such as electric windows and seat warmers.

When you add electric elements to the vehicle you are cutting into that one-third buffer. Depending upon how deeply you cut, eventually you'll be driving one cold night with your high beams on, and your headlights will be start to dim and your radio won't play and if you keep on driving you'll flatten your battery because you are producing less power than you are consuming.

Even if you never have that kind of charging problem, extra load on your alternator means that it has to work harder, on average, than what it was designed for. This will shorten the life of the alternator.

Electrical Load and Fuel Economy

Additional electrical loads will reduce miles per gallon, though not by much. A vegetable oil conversion system that draws every bit of a very large 500 watts would only require an additional gallon of fuel for every 11 hours or so of operation.

Spot Electric Heating

Electric elements are often the best way to apply heat to difficult areas like large filters or pumps. In addition, electric heat can be used effectively in conjunction with an efficient coolant-based system to boost vegetable oil temperatures. By letting coolant do most of the heating, from say 70°F to 160°F, the electric elements can heat the oil well above 160°F without having to draw nearly as much amperage.

Hose-on-Hose vs Hose-in-Hose

Another popular strategy for heating fuel lines is called hose-on-hose. Hose-on-hose just means that a fuel line runs right next to a coolant hose and they are held against each other with zip-tie or other mechanical means. Hose-on-hose is much less efficient than hose-in-hose at transferring heat from the coolant to the fuel line, since the only area of heat exchange is the small areas where the fuel line and coolant hose touch. However, hose-on-hose lines means that the fuel line can be a flexible hose instead of tubing, which makes this option less expensive and somewhat easier to work with.

Tank Heater

In some climates and using some oils, no heat may be necessary in the tank. However in most temperate climates some heat will be necessary in order to easily draw the oil into the fuel lines throughout the year. This tank heating is usually accomplished by a submerged coolant. This can be a simple up and down loop, with very little surface area, or a fluted coil with a great deal more. Electric pad warmers have been used as well. How much heat you need really depends upon your climate. We built the Frybrid In-Tank Heater to handle every climate, but if you are building your own you may not need that level of flexibility.

Rules of Thumb for Heat Exchange

Heat exchange is very complicated area of engineering, but there are few good rules of thumb.

The first rule is that the *amount* of heat exchange that occurs between two liquids is directly related to the amount of surface area shared by the liquids. The second rule is that the *rate* of heat exchange is directly related to the difference in temperature between the two fluids. And the third rule is that counterflow systems are much more efficient than parallel-flow systems (Fig. 3.2).

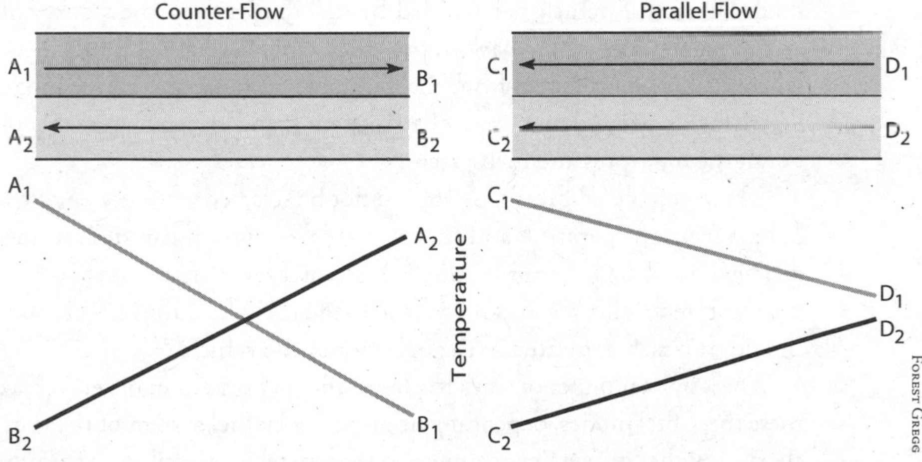

Fig. 3.2: *Counterflow systems are more efficient than parallel-flow systems. Surface area and temperature differential affect performance as well.*

Sizing Your Alternator

When choosing the alternator you need, use the following formula:

$$Amp\ Rating = 1.5 \cdot \frac{Continuous\ Power + Prolonged\ Power + (Intermittent\ Power \cdot 0.1)}{14\ Volts}$$

- Where *Continuous Power* is the watts drawn by components on constantly: instruments, fuel injection, electric lift pump, etc.,
- Where *Prolonged Power* is the watts drawn by components on for extended periods of time: side and tail lights, headlights, radio, dash lights, etc.
- And where *Intermittent Power* is the watts drawn by components that are only on for short periods of time: brake lights, heater, horn, window motors, etc.

Or you can use this formula:

$$Upgraded\ Rating = Stock\ Rating + 1.5 \cdot \frac{Watt\ draw\ of\ new\ Components}{14}$$

Controlling the Fuel Source

In a two-tank system, the vehicle should have three fuel modes: stock fuel mode, where the vehicle is being fed by fuel drawn from the stock tank and returning fuel is sent back to the stock tank; vegetable oil fuel mode where fuel is being drawn from the vegetable oil tank, and purge mode, where fuel is being drawn from the stock tank and clearing vegetable oil out of the high-pressure fuel system.

The vegetable oil fuel mode should not be selected until the engine is at operating temperature and the vegetable oil system can deliver adequately heated fuel. In our opinion, this switching of fuel modes is best done automatically, but it can be done manually. The current fuel mode should be clearly indicated to the operator of the vehicle.

There are a number of ways to plumb the fuel system in order to have these three fuel modes, depending upon the stock fuel system of the vehicle, the number of fuel lines running to the vegetable oil tank, and the type and number of valves. We recommend having both a feed and return line for the vegetable oil fuel, and at least two valves.

Plumbing

Application of Principles

In the design of the plumbing, two principles are really key:

1. Vegetable oil fuel should not contaminate the diesel fuel supply.
2. Purge time, the time it takes for the stock fuel to clear out the vegetable oil fuel at shutdown, should be minimized.

Plumbing with Two 3-Port Valves and Vegetable Oil Return Line

This configuration was popularized by Frybrid (Fig. 3.3). At the time it was introduced, other systems plumbed the vegetable oil tank with only a feed line, some had no purge cycle at all, while others purged the system backwards through the feed line and filter. The Frybrid system employs a return line to the vegetable oil tank as well as a feed line. With this configuration, we were able to reduce purge times without forcing fuel backwards through the filter. This not only prevents possible damage to the filter but allows any air accumulated in the system to be dumped back to the tank rather than simply forced back into the feed line.

Several available systems utilized no purge cycle, which meant that in some systems when fuel mode switched back to the stock fuel, any vegetable oil in the system was being returned to the diesel tank, contaminating it. Other systems simply eliminated the return lines to both tanks meaning

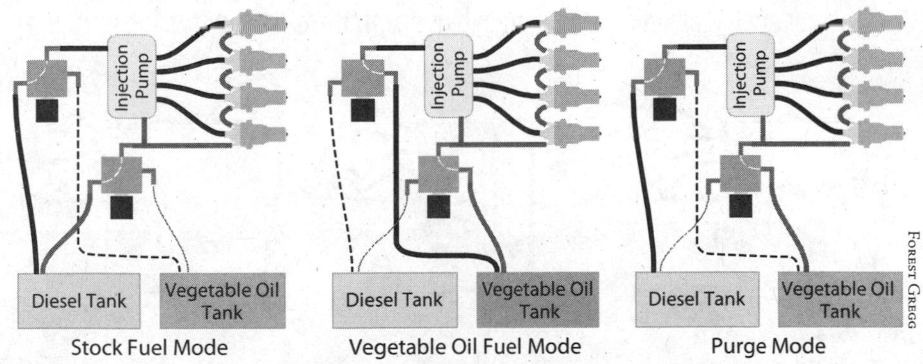

Fig. 3.3: *Plumbing with Two 3-Port Valves and Vegetable Oil Return Line. This system employs a return line to the vegetable oil tank as well as a feed line.*

that the vegetable oil in the system had to be consumed several times over before the fuel in the system was completely replaced with diesel, and this resulted in purge times in the 5 to 15-minute range as opposed to the 10 to 20-second range accomplished by utilizing a return line to the vegetable oil tank. In addition these "permanently looped" systems suffer the adverse effects of overheating the diesel fuel, causing a loss of lubricity and power.

Plumbing with Three 3-Port Valves and Vegetable Oil Return Line, Controlled Loop

This is a recent evolution of the plumbing configuration in Fig. 3.3. We became concerned with returning vegetable oil raising the temperature of the vegetable oil fuel tank and accelerating the reaction of oxidative polymerization. By looping the return — sending the fuel bypassed by the injection pump or high-pressure pump, and injectors back to the inlet side of the fuel system instead of all the way back to the tank — we can dramatically lower tank temperatures (Fig. 3.4).

Looping the return has a second advantage of reducing the amount of oil that must be pulled by the lift pump or injection pump all the way from the tank, reducing vacuum, and vacuum leaks. The downside is that it requires a third valve.

Earlier Designs and Their Problems

Many two-tank systems use only one fuel line for vegetable oil fuel, because it makes installations somewhat simpler and cheaper. However, it also

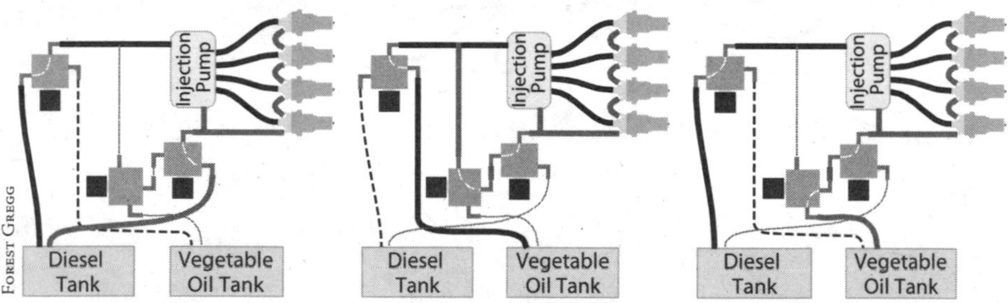

Fig. 3.4: *Plumbing with Three 3-Port Valves and Vegetable Oil Return Line, Controlled Loop. Looping the return lowers tank temperatures.*

means that that the fuel modes will have various deficiencies, no matter how complicated the plumbing.

Fuel Modes with One 3-Port Valve, Hard Plumbed Looped Return

With this setup there are only two modes, stock fuel or vegetable oil fuel, and the returning fuel is always looped. With this setup the diesel fuel can overheat as it loops, which will reduce lubricity and cause wear. The loop also means that it can take a long time for vegetable oil fuel to be cleared from the high-pressure fuel system prior to shutdown, since the only thing taking fuel out of the loop is fuel being sprayed through the injectors. Air leaks are a problem in this design, as there is no way for air to be removed if it enters the loop except by passing through the injectors (Fig. 3.5).

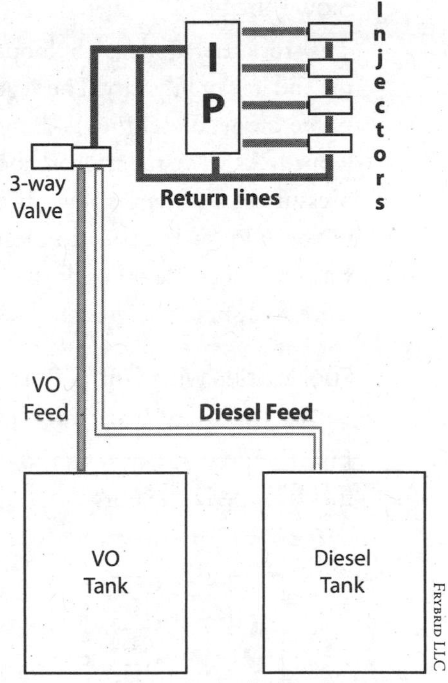

Fig. 3.5: *Fuel Modes with One 3-Port Valve, Hard Plumbed Looped Return.*

Fuel Modes with Two 3-Port Valves, Default Looped Return, Backfeed Purge

This design adds a purge mode, so it takes much less time to clean vegetable oil out of the fuel system prior to shutdown. In purge mode, fuel is being drawn from the stock tank, and returning fuel from the high-pressure fuel system is backfed down the vegetable oil supply line. The diesel fuel is still looped, potentially leading to problems associated with loss of lubricity. If air builds up in the loop, the system can be switched to purge mode, however the purge must be held for an underdetermined length of time to make sure that all the air bubbles are driven all the way though the vegetable oil supply (Fig. 3.6).

Fuel Modes with Two 3-Port Valves, Looped Vegetable Oil Return, Slow Purge

The stock fuel mode is not looped, which avoids problems with overheating and loss of lubricity. The vegetable oil mode has a looped return. In the purge mode, diesel fuel is drawn into the fuel system and the return is looped. As we've mentioned above, this method of cleaning out the high pressure fuel system is slow. If air builds up in the loop, it can be quickly removed by switch back to diesel mode, but this has the consequence of washing vegetable oil back to the stock fuel tank and contaminating the stock fuel supply (Fig. 3.7).

Fuel Modes with One 6-Port valve and One 3-Port valve, Looped Vegetable Oil Return, Backfeed Purge

Like the previous system, the stock fuel is not looped and vegetable oil fuel is looped. The purge is accomplished by backfeeding returning stock

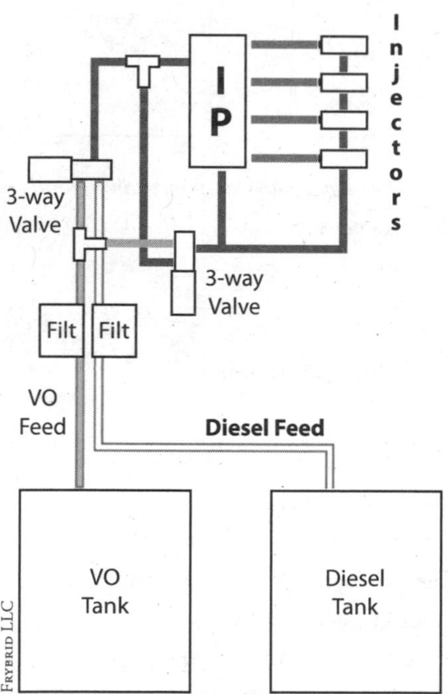

Fig. 3.6: *Fuel Modes with Two 3-Port Valves, Default Looped Return, Backfeed Purge.*

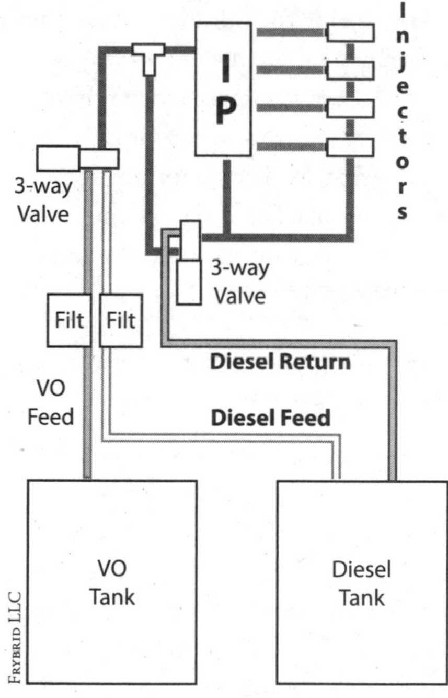

Fig. 3.7: *Fuel Modes with Two 3-Port Valves, Looped Vegetable Oil Return, Slow Purge.*

fuel back down the vegetable oil supply line, which is a fairly fast method. However, if purging to remove air that has built up in the fuel loop, the purge must be held down for a prolonged period of time to make sure that the air is completely carried all the way back down the vegetable oil supply line to the vegetable oil tank.

Notice in this diagram, in vegetable oil fuel mode, that the returning vegetable oil fuel is looped back before the filter. This is usually a poor idea, since one of the main advantages of looping vegetable oil fuel is lowering the amount of pressure or vacuum necessary to move the fuel through the system. The point of greatest obstruction in the flow of the fuel is the fuel filter, and by looping before the filter, most of this advantage of looping is lost (Fig. 3.8).

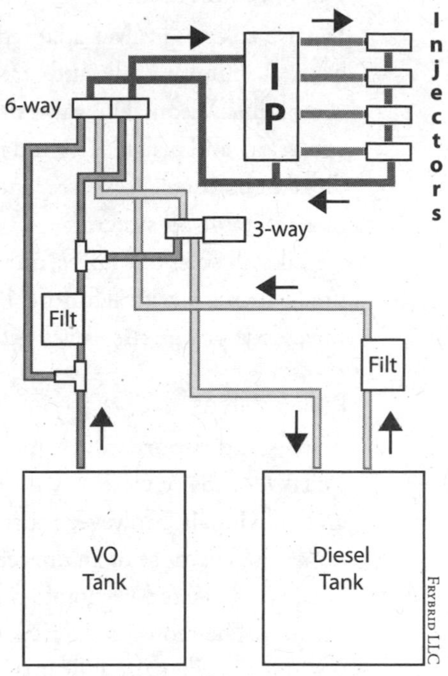

Fig. 3.8: *Fuel Modes with One 6-Port valve and One 3-Port valve, Looped Vegetable Oil Return, Backfeed Purge.*

Valves

Application of Principles

When selecting what valves to use, these principles are important.

1. Materials that are incompatible or reactive with vegetable oil should not be used.
2. Likely points of failure should fail gracefully and obviously.

Flow rate may also be of concern, though most valves sold for vegetable oil have adequate flow rate for most applications.

To flesh it out: valves should be made of material compatible with diesel, biodiesel, and vegetable oil, be able to handle the temperatures of adequately heated vegetable oil fuel, have an adequate flow rate, not draw excessive power, and should have a default mode.

Our Frybrid Valves

We use solenoid valves adapted from the hydraulic industry that consist of an aluminum body, and stainless steel plunger. All seals are biodiesel compatible Viton. The valve draws approximately 1 amp during normal operation, and provides a good flow rate. The valves were designed to handle temperatures, pressures, and vacuums much in excess of those typical of a vegetable oil system.

Like all solenoid valves, our valves have a default or unenergized position, which is a critical feature. If something goes wrong electrically and the valves lose power, the valves default to stock fuel position.

Pollak Valves

Three- and 6-port valves, motorized and solenoid types, made by the Pollak Company, were very popular, and are still used by some companies and individuals. However, most of the field has moved away from the solenoid valves because of an unacceptably high rate of failure when used with adequately heated vegetable oil or in systems that had high pressure or vacuum. The motorized Pollak valves are more robust, but have no default position. If the valve fails it stays in the fuel mode that it was in when it failed. Plantdrive still sells a motorized 6-Port Pollak valve and warranties these valves for one year, although they do not recommend this valve for all applications.

Manual Valves

For the person who has to do everything themselves, manual valves can work well. Two-way ball valves are available inexpensively at any hardware or plumbing store. McMaster-Carr sells a number of interesting 3-way ball valves, though they are no cheaper than a high-quality solenoid valve. While manual ball valves certainly have an advantage in robustness, they have three disadvantages: (1) They usually must be mounted inside the passenger compartment which means that both the stock fuel and vegetable oil fuel supply and return lines must also run inside the passenger compartment, (2) Because they must be accessible to the driver, there is necessarily quite a distance between the valves and the inlet of the high pressure fuel system which can lead to heat loss, and (3) Lastly, the switching of manual ball valves can obviously not be automated.

Cable-actuated ball valves avoid the first of these two disadvantages, but are not cheap.

Control Method

In most two-tank systems, the driver must manually control fuel modes by toggling switches. Although many people wouldn't give up control for anything, user control does mean user errors. People switch over too early, before the engine or oil is hot enough, leading to incomplete combustion lube oil contamination; they forget to purge and the car won't start up the next morning because there is Crisco in the fuel lines; or they forget to switch out of purge mode and empty their stock fuel into the vegetable oil tank perhaps dumping gallons of fuel on the road.

Driving is a complicated enough task as it is, so we use a microprocessor controller to completely automate the switching of most of the fuel modes. In addition to making most user errors impossible and greatly reducing the risk of the rest, automation allows us to switch from the stock fuel to vegetable oil much more precisely and allows us to reliably and safely use a very fast purge mode.

Knowing When to Switch

Whatever controls the switching, vegetable oil mode should not be selected until the engine is at operating temperature and the oil can be guaranteed to be heated to minimum injection temperature.

Coolant temperature gives the best indication of when the engine is at operating temperature, and in systems that use coolant to heat the vegetable oil it is also the best indicator of whether the fuel will be heated to an appropriate temperature, when relying on coolant heat to adequately heat the vegetable oil.

Using coolant temperature to ensure adequate vegetable oil temperature may seem strange. The key is to design the system to have enough heat exchange so that if you know the inlet coolant temperature you can know the outlet vegetable oil fuel temperature with a high degree of certainty.

We did a great deal of testing when designing the Frybrid system, including putting a fuel tank in a freezer until the vegetable oil fuel was a solid block of fat, and we found that even at these sub-freezing temperatures, once the engine coolant reached 180°F, our heated system would deliver at least 160°F vegetable oil.

Measuring the vegetable oil temperature seems like it would make sense, and that information can be useful, but measuring the temperature of a fluid that is not flowing cannot reliably indicate what temperature the fuel will reach when it is moving. Since, in most systems, the vegetable oil fuel is not flowing during stock fuel mode, measuring the temperature of the vegetable oil fuel while operating on the stock fuel does not provide any information that can be easily related to the temperature of the fuel once the oil begins to flow.

Manual vs Automatic Switching

We are big proponents of automated switching between fuel modes. For the switch from stock fuel mode to vegetable oil fuel mode, automated switching has the advantage of not requiring the driver to take his or her attention away from driving to watch a temperature gauge and toggle switches or turn valves. We've also found that when you make it easy for people to switch from stock fuel to vegetable oil, most people cannot resist switching early, before the engine is up to temperature and before adequately heated oil can be ensured. In our system, a coolant temperature sensor sends a signal to a microprocessor controller which switches the fuel as soon as the system can provide adequately heated oil to an adequately warmed up engine.

The other big advantage of automatic switching is being able to have a purge mode with reliable delayed switching. The fastest way to purge with a two-valve system is to feed the engine with the stock fuel and send the returning fuel back to the vegetable oil tank. This method will typically clear the high-pressure fuel system of vegetable oil in 10 - 20 seconds depending upon the engine. In order to do this manually you would need to switch the feed valve, wait the right amount of time, and then switch the return valve. Waiting too long will result in more stock fuel than necessary going to the vegetable oil tank, which will ultimately mean that you will use more stock fuel. Not waiting long enough means that you will contaminate the stock fuel tank with vegetable oil.

Purging is usually best done right before the vehicle reaches its destination, which is often a time that requires heightened attention to driving, and is not the ideal time to be keeping track of time and position of valves. With our system, when you install the conversion system, there's a step to determine how long the purge takes on your system, and where you set a timer that controls the length of the purge fuel mode. After installation, when the time comes to purge, you push a button and the purge mode runs its course automatically.

Mode Indicators

The conversion system should provide clear indication to the driver about what fuel mode the vehicle is running on, and there should be an unignorable signal to the driver if the vehicle is shut down without purging.

The Frybrid system uses three LED lights to indicate the fuel mode. When the vehicle is running on the stock fuel, a red light for the stock fuel mode, a green light for vegetable oil fuel mode, and a yellow light when purging. The signals that control these LEDs could be used to control any arbitrary electronic device.

In our system, if you shut down your vehicle without purging, a loud buzzer will come on until you restart the vehicle and switch to purge mode. We think that this is such an important feature that we have supplied a simple relay-based wiring diagram for do-it-yourselfers who want this feature (Fig. 3.9).

Manual Switching

If you want to forgo the simplicity and robustness of an automatic system, it is critical that you install an accurate coolant temperature sensor and gauge. The stock coolant temperature gauge in a vehicle's instrument

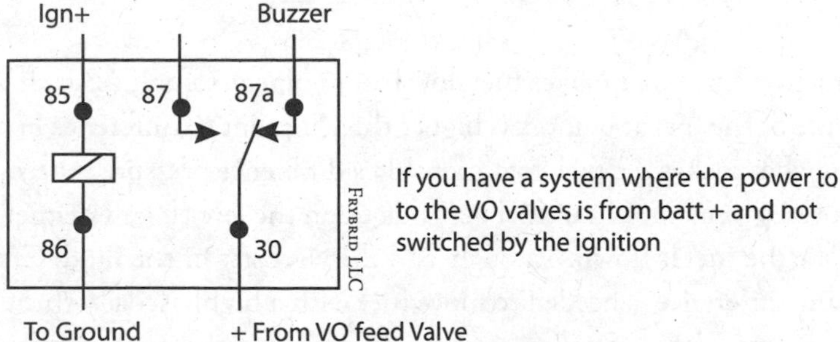

Fig. 3.9: *A relay-based wiring diagram for the do-it-yourself purge alarm.*

panel is *not* accurate enough to now when to switch successfully. We also strongly encourage the installation of a warning buzzer to avoid turning the vehicle off without purging.

Controllers

Frybrid does not currently sell our controller as a stand-alone unit. You can buy a controller from VO Control Systems or Smartveg, and while I think these offerings are somewhat over-priced and over-complicated, they do work.

Filters and Restrictions

Filters should be selected that are easy to heat, have a large surface area, and are at least rated to the same micron level as the stock fuel filter.

Micron Rating

The fuel filter is the last line of defense in protecting your engine from particles that will damage the high pressure fuel system. It should not be used as part of your filtering process, but only as a safeguard for contaminated fuel. Fuel filters are typically rated by micron, a unit less than one-fiftieth the size of a human hair.[1] For instance, a 5-micron filter will stop most particles larger than five microns. What percentage of particles will be removed depends upon whether the rating is absolute or nominal. For a spin-on or cartridge filter, an absolute rating typically means that the filter will catch over 98.7 percent of particles larger than the micron rating. According to fuel providers and filter makers, particles larger than five microns represent a danger of increased wear in the high-pressure fuel system. High injection pressures of newer diesel engines increase the hazard of smaller particles.

Head Loss

In a fuel system, what makes fuel flow from point A to point B is that the pressure of the fuel at point A is higher than at point B: difference in pressure is what makes the fuel flow. How big a difference pressure is necessary to move a given amount of fuel depends upon the length and diameter of the tube the fuel is flowing though, and the viscosity of the fluid. Greater pressure difference is needed to move fuel with a high viscosity through a long, skinny tube. A smaller pressure difference is required to move the same amount of less viscous fuel though a short, wide tube.

The total pressure difference needed to provide an adequate volume of fuel is all the pressure difference necessary to move enough fuel across each individual component of the fuel system, all added together. When you split a fuel system into individual components, the fuel filter is generally the part of the fuel system that requires the greatest pressure difference (Fig. 3.10).

An improperly chosen, heated, or clogged filter will require a very large difference to move fuel through it, which means the pump will have to work harder to produce either higher vacuum or higher pressure, depending upon whether the filter is on the inlet or outlet side of the pump (Fig. 3.11). If the pump is not equal to the task the restriction will cause the engine to starve of fuel.

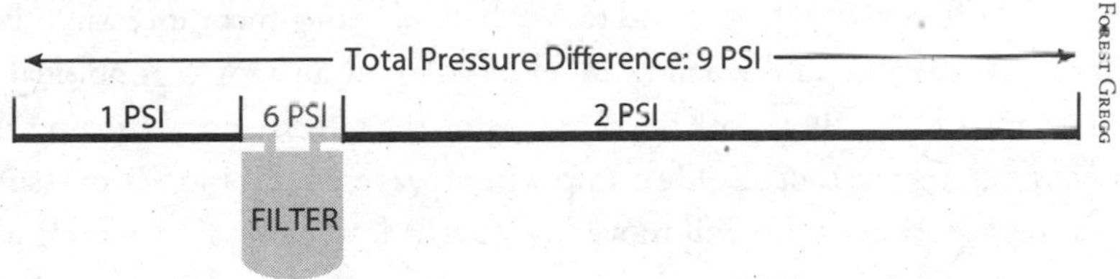

Fig. 3.10: *The fuel filter is the single fuel system component that requires the greatest pressure difference.*

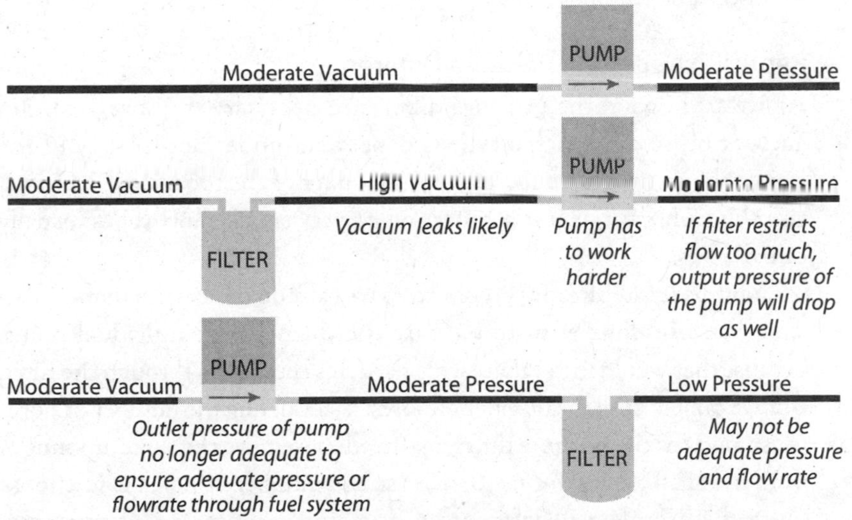

Fig. 3.11: *Filter problems will cause the pump to work harder, whether it is drawing a vacuum or delivering pressure.*

Vacuum Leaks

When a fuel system is under vacuum, air will try to push its way into the fuel system anywhere it can, typically at connections. Vacuum leaks can be very difficult to track down and fix because there is no visual indicator of where exactly the leak is taking place and because it is very difficult to make a perfect and stable air-tight seal. If you do have a vacuum leak, then you have two problems. The first is that the air reduces the level of vacuum at the point where it infiltrates the fuel system, which means that there is less vacuum to pull fuel from upstream, which will reduce fuel flow and can cause your engine to starve.

The second problem is that air in the fuel makes pumps work very inefficiently. Fuel pumps depend upon the fuel being uncompressible, but air is compressible, and when the fuel has a lot of air in it, a good deal of the action of the pump goes to compressing the air instead of moving the fuel, which means that the lift pump will move less fuel at lower pressure, and the injection pump will move less fuel at lower pressure, which will reduce the pressure at the injectors and cause a poor spray pattern and incomplete combustion. In a worst case scenario, air in the fuel lines will cause an air lock, where the pump is just compressing and decompressing air instead of pumping fuel.

Reducing Necessary Pressure Differences

As we said above, the pressure difference necessary to move a certain amount of fluid through a tube is dependent upon the viscosity of the fluid, the length of the tube, and the diameter of the tube. Well, a filter, if you think about it, is just a collection of very small, short tubes, usually called pores.

Now pores are already very short, so we can't do much with that dimension. We also don't want to increase the diameter of individual pores, because that would mean that larger particles could pass through the filter. But we can increase the number of pores. By doubling the number of pores we should cut the pressure difference needed to move the same amount of fluid in half. The way for us to increase the number of pores is to choose filters that have larger surface areas. Sometime that means a larger overall filter, but filter surface area can be increased without increasing total filter size though pleating of the filter material or similar means.

Rather than overall physical size of the filter, a better indicator of surface area is flow rate, and most filter manufacturers provide information on flow rates at different pressures. If two filters have the same micron rating, the one with the greater flow rate at the same pressure will be the one that requires less of a pressure difference in your fuel system. You can also increase the surface area and cut the necessary pressure difference in half by using two identical filters plumbed in parallel.

Heating

Heating the oil and reducing the viscosity also makes it easier for the fluid to flow through the filter and will reduce the necessary pressure difference. Heating the oil from 70°F to 100°F will cut the viscosity in half and will also cut the pressure difference needed to move it through the filter — and the entire fuel system, for that matter — by half. The oil should already be heated by the time that the oil enters the filter, but the filter should also be heated to prevent high-melting-point fats from clogging the pores, and also because filters are often sites of great heat loss.

Not much heat is really necessary for the secondary functions and an electric wrap or coiled coolant loop is likely sufficient. We use a small flat-plate heat exchanger for double duty. Oil is passed through the heat exchanger before entering the filter, significantly boosting the temperature of the oil and reducing viscosity, and since the heat exchanger is mechanically connected to the filter, it is effective at heating the filter body itself and melting any high-melting-point fats before the vehicle is switched to vegetable oil fuel.

Other Restrictions

Fuel Lines. Small variations in the diameter of fuel lines can make a big difference in how much pressure difference is necessary to move a given amount of fluid through them. The table below (Table 3.1) compares pressure differences that would be expected by changing the internal diameter of a fuel line, assuming that everything else remains the same and that it

½″	⅝″	¾″	⅞″	1″
50 PSI	20 PSI	10 PSI	5 PSI	3 PSI

Table 3.1: *Pressure Differentials with Change in Diameters.*

takes a difference of 10 psi to move the fluid through a tube with ¾-inch internal diameter.

Valves. We discuss the flow rates of valves in the section on controlling the fuel source.

Pumps

Nearly every diesel engine has a pump to supply fuel to the injection pump or high-pressure pump, usually called the lift pump. A vehicle may need to include an additional electric lift pump as part of the vegetable oil conversion if vehicle does not have a lift pump in the engine bay or if the stock lift pump cannot function adequately when pumping thicker vegetable oil. See the Appendix C for which vehicles will need a pump.

What Pumps Don't Work

Experience has shown us that most electric pumps designed to pump diesel fuel don't work well with vegetable oil. We have tried pumps from Carter, Facet, Walbro, Aeromotive, and Shurflo, and none have performed particularly well. Mallory pumps can be acceptable if they are purged of vegetable oil at shutdown and never have to pump cold vegetable oil. As far as types of pumps, we have not had good luck with vane, solenoid, or diaphragm-style pumps.

What Pumps Do Work

Internal gear pumps seem to be the type of pump best suited to pumping vegetable oil. Diesel Performance Products is selling a gear pump that they warranty for one year with use of vegetable oil, the only pump maker to provide any warranty that covers vegetable oil. As of this writing, this pump and warranty are only available through Frybrid, however these pumps are also sold as part of Diesel Performance Product's F.A.S.S. system, which has distributors across the country. This is the only pump that we use in applications where the pump will always have vegetable oil in it.

We use Mallory gear-style pumps for some relatively low pressure applications — less than 20 psi — but only when vegetable oil will be flushed with diesel at shutdown, otherwise these Mallory pumps do not last very long.

Material Compatability

Appropriate materials must be selected for the vegetable oil fuel system. Copper and steel should be avoided because they are strong accelerants of the oxidative breakdown of vegetable oil. Natural rubber and some other polymeric material should be avoided as vegetable oil will dissolve these materials.

Fuel Tanks

Aluminum has proven to be the best material for vegetable oil fuel tanks. Aluminum is light, weldable, and relatively inexpensive. It is still an accelerant to oxidation, but is the least reactive of common metals. Tanks made of HDXLPE, a high-density cross-linked polyethelyne, are also very promising. They will not react with vegetable oil, are even lighter, and can be much cheaper. Like all plastics, HDXLPE will eventually break down under UV light, so HDXLPE tanks should be protected from sunlight.

We are still experimenting with HDXLPE, but it is the most promising plastic material we know about. All the other plastics we looked at were unacceptable because of inadequate temperature rating, material incompatability, or brittleness.

Steel tanks often appear attractive, because they can be had very cheaply at the local junkyard, however we strongly discourage using them. Steel is a powerful accelerant of oxidative breakdown of vegetable oil and polymerized oil seems to form thick paint-like skins selectively on steel. Stainless steel can be used, however it is expensive, and unless a very high grade is used, it will also create problems with oxidation and polymerization.

Heat Exchangers

Nearly all off-the-shelf coaxial, flat plate, or tube-and-shell heat exchangers contain copper because of the excellent heat transferring properties of the metal. While any exposure of vegetable oil to copper should be avoided, especially when the oil is hot, using a final fuel heat exchanger that does not contain copper or brass (an alloy of copper) usually means much higher expense and lower thermal efficiency.

We do use final heat exchangers that contain copper, but we also loop the returning fuel so that it will be consumed by the engine in short order, instead of going back to the fuel tank where it will be again exposed to air.

Copper tubing is easy to work with and is an excellent thermal conductor which makes it an attractive material to use for in-tank coolant loops. Do not do this. The tank is the place where the oil is at greatest risk of going bad due to oxidative breakdown; there is little point in adding a significant quantity of the most common catalyst to that breakdown. It doesn't matter whether the copper sits at the bottom of the tank, and is usually submerged. It will still speed up the breakdown of the oil. Aluminum tubing is a good substitute.

Brass compression fittings are sometimes used inside a fuel tank. Besides the fact that compression fittings have an unacceptable rate of failure, as we will discuss in the next section, this is a bad idea because brass is an alloy of copper and also an efficient catalyst for oxidative breakdown of vegetable oil.

Fuel Lines

We use aluminum or biodiesel-rated rubber flexible lines. Stainless steel is an acceptable metal, but more expensive and somewhat harder to work with. We use biodiesel rated lines because many of our customers want that option, but any diesel-rated flexible line that is also temperature-rated up to 180°F should be acceptable. To our knowledge, there is no clear, flexible fuel line that will not eventually fail when exposed to hot vegetable oil.

Using PEX tubing is popular because of its low price, but it is not a great solution. Even the highest grade of PEX will still be moderately effected by petrodiesel, and only has a recommended maximum operating temperature of 180°F.

Common Points of Failure

Compression Fittings

Compression fittings are commonly used to make hose-in-hose fuel lines. In some systems they are used as a way to couple tubing to threaded pipe, usually inappropriately.

The typical brass compression fittings you can buy at the local hardware store are designed to connect a tube to a pipe or fixture without soldering. The threaded part of the compression fitting is screwed into the pipe or fixture. The nut and collet slide onto the smooth tube and the tube is inserted into the fitting until it hits the shoulder, then the nut is tightened.

As the nut is tightened, it compresses the brass collet and if all goes well the collet forms a leak-free mechanical seal between the tube and the compression fitting. The typical compression fittings are only intended for applications where there will be no movement at the connection.

Since everything is always vibrating in an automobile, it is not surprising that compression fittings are a common point of failure in vegetable oil conversion systems. The problem is made ever worse by the fact that many of us physically alter the compression fittings to use them in hose-in-hose systems. Specifically we ream out the inside of shoulder of the compression fitting so that the entire fitting will slide over the outside of the tube. If care is not taken, it is very easy to scratch the mating surface of the compression fitting, which will prevent it from ever sealing positively.

It is possible to buy compression fittings without the shoulder, usually called bore-through compression fittings. Swagelok and Parker make fittings, similar to compression fittings, but are slightly different designs that are designed to work in vibrating conditions (Fig. 3.12). Using these types of fittings will improve reliability, but they should still not be used in situations where failure of the fitting will lead to contamination of the fuel by engine coolant.

Compression Fittings in the Tank

While we believe that coolant connections to the in-tank heat exchanger should never be made inside the tank, internal connections made with

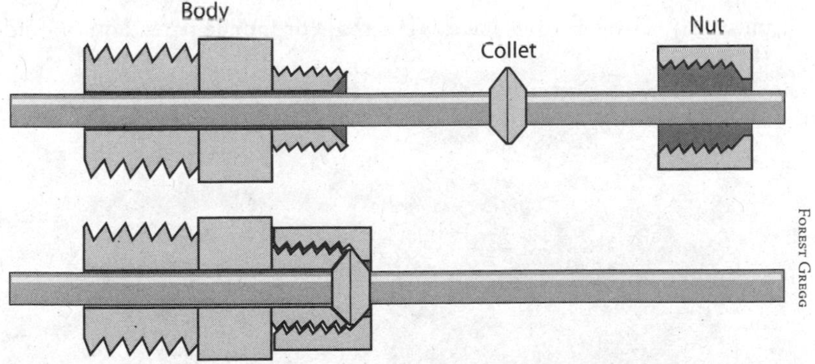

Fig. 3.12: *A bore-through compression fitting, designed to work reliably in vibrating conditions.*

compression fittings are a particularly bad idea. Not only will failure of the fitting lead to coolant contaminating the fuel, but when the failure does occur there is no indication of failure until a great deal of damage has already been done.

Recommendations

We use compression fittings for making hose-in-hose fuel lines, but we are careful to use them in ways that if they fail there is no possibility for the coolant to contaminate the fuel and that if they do fail, the failure is easily noticeable. We are satisfied with using conventional compression fittings that we bore out on a drill press and are carefully checked for scratches on the mating surface.

We have moved away from using brass collets and now use plastic collets. The plastic collets do not necessarily seal any better, but they are much easier to work with if the fitting fails. When using a metal collet, it is usually impossible to remove the collet after the nut has been tightened down. The tube will usually have to be cut and if there is not enough slack in the fuel line, the entire fuel line may have to replaced. Plastic collets can usually be removed easily.

Hose Barbs

Every connection in the fuel system is a potential point of failure for leaks — fuel leaking out if the fuel is under pressure, or air leaking in if the fuel is under vacuum. Most conversion systems use hose barbs for many of the connections. These fittings have NPT male or female thread on one side

Forest Gregg

Fig. 3.13: *A typical hose barb.*

and a barb on the other (Fig. 3.13). Flexible hose slides over the barb and is secured with a hose clamp. Using a good thread sealant and a proper hose clamp will help avoid leaks (we use Permatex Teflon thread sealant 14B and Swedish hose clamps that don't cut into flexible fuel line), but we are moving away from these types of fittings altogether, in favor of JIC and Push-Lok fittings. These type of fittings are extremely reliable in providing a leak-free connection, and do away with the necessity of thread sealant and hose clamps. Unfortunately, these fittings can be hard to find and are usually more expensive than hose barbs.

User Errors

In many systems it is easy for the user to make a number of errors that can have unfortunate consequences. Following are a few examples:

Forgetting to Purge. If the vehicle is shut down without purging the vegetable oil fuel out of the high pressure fuel system, it can be extremely difficult to restart the vehicle. If this does happen, then hot water should be poured over the injectors, injector lines and injector pumps, until they are warm to the touch. If the engine will still not start, it may be necessary to crack an injection line or two from the injector and turn the engine over until the stock fuel starts coming out the injection line.

Filling the Stock Tank with Vegetable Oil or Vice Versa. Depending upon the plumbing of the system it can be possible to pump the fuel from the stock tank into the vegetable oil tank or vice versa. Diesels typically bypass most of the fuel that reaches the inlet of the high pressure fuel system and if this fuel is returned to the wrong tank, it can send all the fuel from one tank to the other in short order.

Recommendations

The Frybrid Controller was specifically designed to make it difficult or impossible for these user errors to occur. A very loud buzzer comes on if the vehicle is shut down without purging, and the way that the valves are controlled make it impossible to pump all the fuel from one tank to the other.

This automated approach is really the only way to really avoid these problems, however, clearly visible indicator lights and buzzers can and should be used to make these user errors less likely in a manually controlled system.

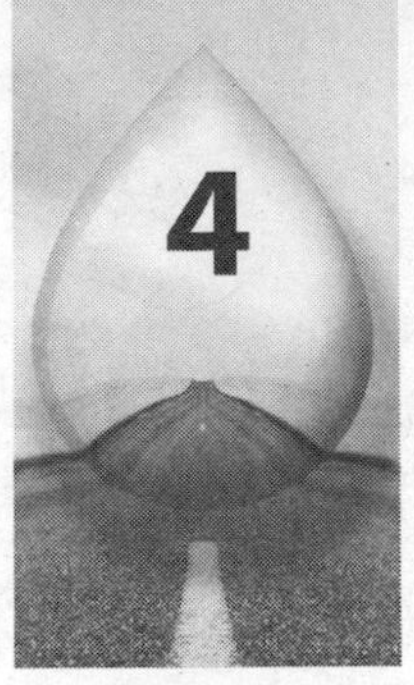

Practical Issues

Up to this point we have been focused on the theory and the engineering that allow us to successfully fuel a conventional diesel engine with straight vegetable oil. However, actually using vegetable oil as a fuel, day to day, requires a proactive understanding that has little to do with the conversion itself. You must find oil, negotiate with restaurants, collect it, clean it to an acceptable level, understand the legal implications of your actions, and perhaps join with others to make these activities less time consuming or expensive.

In this section, we are going to discuss these practical concerns. We'll start by talking about what steps you should take before you convert your vehicle and the additional maintenance that will be needed on your converted vehicle. Then we'll talk about what oil sources you can use, and how to go about collecting oil from restaurants. We will then discuss the legal landscape of vegetable oil and how co-ops might be a good option for you to share the time and expense of collecting and filtering oil.

Preparing and Maintaining a Vehicle for Conversion

If your vehicle is not running well on the fuel it was designed for, it probably won't run any better on an alternative fuel, and depending upon what's causing the engine to perform badly, running vegetable oil will speed the wear. Before you consider converting the vehicle, have a skilled diesel mechanic give your engine a clean bill of health. There are three tests that you must do: a compression test, injector tests, and a test of the cooling system.

There are many reasons why an engine can have poor compression. If it is because the piston rings are worn, then when you are running on vegetable oil an unacceptably high level of vegetable oil fuel will bypass the rings and contaminate the lubricating oil which can lead to catastrophic failure.

As for the injectors, take them out and have them tested and set at a good injection service shop. Only accept injectors that both open and close at the appropriate pop pressure for your engine, and which show a good spray pattern. Do not assume that new injectors will open and close at the appropriate pressure. If the injectors are not set properly, you will either underfuel, which means that the engine will produce less power, or overfuel, which will cause the injector to continue to inject vegetable oil past the point where it can be completely combusted, leading to carbonization and contamination of the lubricating oil.

If the cooling system is not working well, then it will be unable to effectively heat the oil in a conversion that depends upon coolant heating. Make sure that the cooling system is in good working order and is able to maintain operating temperatures.

An excellent practice, considering that this is still a rather experimental endeavor, is to have your lubricating oil analyzed at every oil change. An oil analysis consists of the testing the oil for antifreeze, fuel, oxidative products, acidity, solids, viscosities, and the presence of a number of different metals and other elements. The results of these tests can indicate abnormal levels of fuel dilution, antifreeze, particulate contamination, and abnormal wear of critical engine components, and problems with your lubrication system.

By having the oil analyzed two or three times before converting to vegetable oil, you have established a baseline of normal wear, and can catch serious problems before they happen. Then, when you convert the vehicle, you can see how using vegetable oil fuel impacts the wear and properties of the engine oil, which will allow you to make adjustments, if any are necessary, before a serious problem develops.

Caring for your Converted Vehicle

A healthy engine with an adequate vegetable oil fuel conversion system should not require any additional maintenance, besides changing the vegetable oil fuel filter at the same intervals that you change the stock fuel filter. That said, there are many conversion systems, many diesel engines in poor health, and a great deal of variation in how carefully people prepare the vegetable oil fuel. Below we will discuss the problems that can occur.

Less Power on Vegetable Oil

There should be a small loss of power on vegetable oil of about 5 percent, but that's a smaller difference than most people can notice — and it should only be apparent at the top of a gear. Most of the time loss of power on vegetable oil fuel is a result of not enough fuel getting to the injector, because of a clogged fuel filter or a vacuum leak. If you are constantly clogging the fuel filter, then you need to make changes in how you prepare your fuel.

Smoke on Vegetable Oil

Smoke means that the fuel is not burning completely. This can be caused by many things, some of them very serious that we will discuss below, but most of the time smoke on vegetable oil is caused by vacuum leaks allowing air to become entrained in the fuel. Air in the fuel prevents pumps from effectively pressurizing which ultimately leads to fuel being sprayed from the injector at too low a pressure and in a pattern that leads to incomplete combustion.

EGR Valves

The function of an Exhaust Gas Recirculation (EGR) valve is to recirculate a portion of exhaust gases back into the engine's air intake, in order to lower the production of NO_X gases which are formed at the hottest temperature point of the combustion chamber. By replacing oxygen containing fresh air, exhaust gases lower the peak temperatures of the combustion chamber by reducing the completeness of combustion. In diesels, exhaust gas recirculation also decreases NO_X by reducing the size of the excess oxygen- to-fuel ratio at low load conditions, leaving less oxygen available to form NO_X.

By definition, whenever completeness of combustion is reduced, it results in more incompletely burnt fuel in the form of soot, and, ironically buildup of soot on EGR valves is a major cause for the failure of these components. Some engines have EGR valves that seem to become clogged with soot more often when burning vegetable oil. If the EGR valve is becoming clogged repeatedly, then you should look to how you can increase the completeness of combustion by further heating the fuel, or by adjusting the timing of injection.

Serious Problems

When vegetable oil causes or contributes to the breakdown of an engine there are two main pathways of failure: incompletely combusted fuel leading to carbon deposits on components inside the combustion chamber and associated wear, and uncombusted fuel bypassing the rings of the piston and contaminating the lubricating oil, thereby degrading the effectiveness of the engine oil, and leading to wear because of insufficient lubrication.

Carbonization in the Combustion Chamber

Incompletely burnt fuel in the combustion chamber can lead to buildup of carbon deposits on the injectors, injector needle sticking, coking of the intake and exhaust opening, coking on the exhaust valve stem, and deposits on the upper edge of the compression ring groove. Deposits on the injector will lead to a deteriorated spray pattern, which will lead to even worse combustion, and more soot and carbon deposits. Sticking of the injector needle will cause either underfueling which will lead to power loss, or overfueling which will produce more carbon deposits and soot. Deposits on the valve openings or valve stems will lead to a loss of compression, wear of the valve seats, or possibly a bent valve from a stuck valve getting hit by the piston. Deposits on the upper edge of the compression ring groove will cause greater wear of the piston against the sleeve and will also decrease compression and increase blowby. They can also contribute to catastrophic failure caused by a piston seizing.

Symptoms of engine degradation are loss of power, rough running and idling, and smoke. If these symptoms present themselves, take immediate action.

Contamination of the Engine Oil

Engine oil performs two critical functions — lubricating and cooling moving internal components of the engine. If the engine oil is significantly contaminated by vegetable oil, the oil will thicken, and eventually it will be unable to perform these critical functions. Wear of critical engine components will hasten until catastrophic failure of the engine occurs. Turbochargers are often the first engine component to fail.

In an engine running on vegetable oil, the quality of the engine oil needs to be carefully and frequently checked. It is probably prudent to shorten the oil-change intervals, unless oil analysis indicates that this is unnecessary.

Oil Sources

There are four different types of oil stocks that might be used as vegetable oil fuel: unrefined oil, refined cooking oil, waste cooking oil, and yellow grease — each type representing oil at a different stage in the typical life cycle of cooking oil. Of these oil stocks, only refined vegetable oil and waste cooking oil have proven acceptable. We will discuss these types of oil stocks first.

Refined Cooking Oil

Refined cooking oil, the kind that you buy at the store or from a restaurant supply company, can be used as fuel without further processing. Cooking oil undergoes an enormous amount of technical processing to make it one of the most chemically pure products available to regular consumers. Hydrogenated oils, like creamy liquid shortening, are just as free of contaminants, but the greater thickness of these oils may make them unusable unless the conversion system has adequate heat. Cold pressed, virgin, or extra-virgin cooking oils are not refined and have not been proven acceptable.

Until recently, the cost of refined cooking oil was actually lower than petrodiesel. However, the booming biodiesel market has raised the cost of vegetable oil substantially over the past year, and with the continuing growth and acceptance of biodiesel, the price of cooking oil and the price of petrodiesel are now probably permanently coupled.

Waste Cooking Oil

Waste cooking oil is oil that has been used by a restaurant and requires removal of contaminants before it is acceptable as fuel. The vast majority of the field is using waste oil as their fuel source because it is by far the cheapest oil. Depending upon the setup, the total cost of using waste vegetable oil can be pennies on the gallon. At present, most restaurants still have to pay a company to take their waste cooking oil away, but in some cities, such as Seattle, Washington and Asheville, North Carolina, it may be hard to find a restaurant willing to give you their oil for free, as biodiesel refineries have secured contracts with many of the desirable restaurants.

We will discuss how to find good waste oil and how to secure the restaurant as an oil source in the next section.

Unrefined Vegetable Oil

Crude, unrefined vegetable oil has generally proven to be unacceptable as a fuel, owing to the large amount of contaminants present in the unprocessed oil. The only well-documented case that I know of successful long term use of unrefined vegetable oil is that of Finnish mustard farmer Vaino Laiho, who fueled his 1985 Valmet 505 tractor for ten years and over 1,900 test hours on a blend of cold pressed mustard oil and petrodiesel, before the engine finally failed.[1] Laiho screw-pressed the mustard oil, and let it sit for at least a month to allow contaminants to settle. He then drew off the top clear layer of oil and mixed it with diesel in a 60 percent mustard oil, 40 percent petrodiesel blend. The performance of the tractor did not appreciably deteriorate until the tenth year, when the motor oil was changed to a cheaper blend and mustard oil that had been stored for over five years was used in the blend. The motor failed catastrophically within 100 hours, with piston number three seizing and the turbocharger breaking. The cause of the piston seizing was attributed to a brown, gummy substance causing the injectors to stick open.

In addition to the experience of Vaino Laiho, stories of farmers blending unrefined vegetable oil up to 50 percent and reporting good results are fairly common. However, such anecdotal evidence needs to interpreted with caution as stories of success tend to be repeated much more loudly and frequently than failures. Diesel engines for tractors tend to be built much more robustly and designed for greater ease of service than automotive diesels, and it is very difficult to extend reports of success in these types of engines to diesel engines in general.

That said, a sample size of one is much larger than a sample size of none, and it must be conceded that Laiho and his Turgu Polytechnic collaborators have demonstrated that it is possible to use unrefined vegetable oil as fuel for at least one decade without causing serious harm to the engine. Steven Hobbs, an Australian farmer and businessman, is planning on using unrefined vegetable oil on a large scale. His experience may well settle the viability of this oil source.

Yellow Grease

Free Fatty Acids	15% max
Moisture and Impurities	2%
Iodine Value	75-80
[Source: rothsay.ca/specs/yg_spec.html]	

Table 4.1: *A typical analysis of yellow grease.*

The dumpsters behind restaurants are typically owned by rendering companies which take the waste oil, mix it with other waste streams, clean it up somewhat, and sell it as a commodity called yellow grease. For people with big plans, yellow grease can be attractive because you can buy it by the tanker load, you don't have to go through the dirty and time-intensive work of visiting the restauraunts to pick up the oil, and until recently it was very cheap, less than $0.50 a gallon in some places. Unfortunately, as a rule, yellow grease is fairly degraded product. Table 4.1 shows a fairly typical specification of yellow grease.

As we've discussed in the section on "Reactions in the Fryer" free fatty acids and the mono/diglycerides they accompany are very difficult to reduce cheaply and effectively. The Iodine Value of yellow grease should not be used as an indicator of oxidative stability, as it can be used for fresh oil, but instead as an indicator of oxidative degradation[2] and as evidence of the relatively large percentage of saturated fats that come from animal sources that makes yellow grease very thick, even semisolid at room temperature. As far as I know, no one has been able to clean up yellow grease enough to use it as fuel that does not produce problems in short order. Albuquerque Alternative Energies of Albuquerque, New Mexico has been working intensively on this problem, and claim to have found a solution, although, as of this writing, they have not yet released data or details for evaluation.

The great attraction of yellow grease — its price — has largely disappeared, as biodiesel refineries have come on line use yellow grease as a feedstock. As of this writing, the price of yellow grease was over $2.20 a gallon.

Waste Vegetable Oil

Waste vegetable oil is usually the most attractive fuel source for a vegetable oil conversion. If the you find the right sources and set up a good system, it can mean that you are getting fuel for almost no cost.

Finding Good Sources

Depending on how a restaurant uses cooking oil, the waste oil will vary significantly. Choosing good sources will make every step that comes after significantly simpler. The best restaurants are ones in which most of the oil is not used in deep frying. The oil is used once on the preparation of one dish and then discarded, and as a result the oil is very lightly used. Asian restaurants tend to fit this profile.

Second best are restaurants in which the oil is used mainly to deep-fry vegetables, French fries or tempura, for example. Restaurants that do a lot of deep-frying of animal products such as chicken, fish, or beef tend to have the hardest oil to work with, since the oil will contain a lot of animal fat, which will increase the viscosity of the oil, and animal proteins, which will reduce the stability of the oil. Oil used to deep-fry meat will also tend to include a good deal of flour from breading, which forms a gravy in the oil that can be difficult to remove.

Once you have identified a restaurant that you might want to use as a source, go around back and check out the renderer's dumpster. If it makes you gag, you should probably pass. If it's not terrible, ask the restaurant staff how frequently they change their oil and ask for a sample of their waste oil. If they change the oil less than once a week or the smell of the oil is offensive, you should probably pass. If the oil still seems acceptable, take the sample home and let it sit for a couple of days. Much of the particulate will settle out, and will give you an idea of how much material you will have to remove when you clean the oil. If the oil never forms a clear layer, you should probably pass. A more scientific test of oil quality is to measure free fatty acids. A free fatty acid level of 1 to 2 percent usually indicates the oil was not abused. Higher levels would raise concern.

The final test of the quality of the oil is, of course, how easy it is to clean, and sometimes it may be easier to accept oil of unknown quality and then stop accepting the oil if it does not meet your standards than it is to do all the research on the front end.

Color

Color is not a good indicator of the quality of the oil. While it is true that oil darkens as it degrades in the fryer, the color can also be strongly affected by what food was cooked in the oil.

Hydrogenated Oil

If you come across a restaurant with really thick and cloudy waste oil, you should not necessarily pass on it. Ask the restaurant staff if it is hydrogenated. If they don't know, ask to look at the jugs that the oil came in. If the oil says something like "Creamy Liquid Shortening" it is hydrogenated. Hydrogenated oil is an excellent fuel for systems with adequate heat. If you don't have the ability to put a lot of heat in the tank, you should probably pass on this type of oil.

Securing a Source

Once you find a source that you want to pick up from, you need to do two things. You need to make sure that you can collect the oil legally, by figuring out the relationship that the restaurant has with their existing oil collector and take steps to make sure that you can collect the oil without breaking the law, and you need to arrange another way for them to discard the oil. The best setup is to convince the restaurant staff to put the oil back into the containers they came from. This is pretty clean and the jugs can be put in trunk of a car, whereas most other setups require that you have access to a truck. Reusing the original containers is also attractive since it seals the oil from exposure to the air while it awaits your pick up. Restaurants may be resistant to discarding of their oil this way since it means that the oil cannot be discarded hot. Hot frying oil will melt right through the plastic jugs and buckets.

If you can't get a restaurant to reuse the original containers, than you will need to supply them with some other container. Sealable metal pails are a good option, since they allow the restaurant to change the oil hot, can still be put into the trunk of a car.

A common setup is to supply the restaurant with a 55-gallon drum. If you go this route, you will need access to a truck and will need to buy a good 12-volt DC pump, and you will also need some way to protect the oil from the elements. A good way to protect the oil is to build an awning over the barrel out of scrap wood and a plastic tarp. An advantage to using a 55-gallon drum is that you don't need to visit the restaurant as frequently to pick up their oil.

Contracts

If you can, try to get your arrangement with the restaurant in writing. Competition for waste oil is growing, and locking in sources is going to be increasingly important. However, restauranteering is by and large a cash-based informal business, so don't be surprised if the restaurant doesn't want to sign anything.

Restaurant Education

Many restaurants clean out the fryers with soap and water and then discard the wash water with the rest of the oil. Ask the restaurant to put the wash water in a separate container and take this mess with you when you pick up the oil.

Keeping Restaurants Happy

Restaurants are generally very happy to have somebody take their waste oil away for free, especially for such a good use, and it's pretty easy to keep them happy. Be careful not to make a mess and come prepared with supplies to thoroughly clean up any mess that you do make. Give them cat litter or other cleaning supplies to help them clean up any messes they might make when they are putting the oil into containers for you. Show up when you say you will and make sure that the restaurant has a way of getting a hold of you in case they have oil backing up.

Restaurant owners tend to know each other, so a bad experience at one restaurant can make it much more difficult for you to find another restaurant willing to work with you, or anybody else for that matter. By the same token, if you act professionally, word will spread and other restaurants will soon start offering their oil up to you, which is the position you want to be in.

Rendering Companies

Rendering companies collect the waste oil from restaurants, process it, and resell it as a commodity called yellow grease, discussed earlier. Whenever the price of yellow grease has been high, theft has been a problem, and the problem has grown significantly over the past few years with growth of biodiesel and straight vegetable oil conversions. The price of yellow grease has nearly tripled over the past two years.

As prices and competition increase, it is likely that more states will pass legislation similar to a law already on the books in California that makes it a misdemeanor for an individual or company to transport "inedible kitchen grease" without registering with the California Department of Food and Agriculture

As an individual, you can't do much to change the larger economic forces at play, but you can do your part to prevent our field from getting a bad reputation as a bunch of thieves. Simply put, don't steal oil. Even if you have permission from a restaurant, don't take oil from a rendering company's dumpster.

There have been scattered reports of rendering companies using intimidating and illegal tactics to protect their oil sources and even reports of rendering companies stealing oil. If you act professionally, you will have the law on your side and will be in a much better position to protect your sources.

The Law

Your state or local municipality may have laws that control who can collect and transport waste oil. Investigate these laws, and comply with them. Generally, if you are collecting oil for personal use and you are not planning on selling the oil or using the oil in a commercial enterprise, it is inexpensive to comply with regulations.

Processing Waste Cooking Oil

Before your waste cooking oil can be used for fuel it must be cleaned of particulate contaminants and water. There are a number of ways to do this, but the most common methods are settling and filtering, and centrifuge methods.

Before I get into that, I'm going to address some common and incorrect notions about processing waste oil.

Myths

You can just pump it straight into your vehicle from the dumpster.

I actually know a few people who have done this and their vehicle didn't have any trouble for thousands of miles. They just kept on changing their filter every fifty miles. Eventually, they had to drop their tank and clean it out because of a buildup of sediment. Doing this will, at some point, cause long-term damage to your engine. It is safe to assume that waste oil, particularly waste oil that has been sitting around outside, contains a significant amount of water that will cause a number of problems, as we've already discussed. It is also important to understand that your automotive filter will not remove all the particles that the filter is rated to remove. If you are constantly changing your filter that means that a good number of particles are going to get past the filter and eventually cause some expensive damage.

Petrodiesel water block/separating filters remove water from vegetable oil.

Diesel fuel filters commonly have either a water-blocking component or a water separating component. These water-blocking or separating techniques can be based on any number of principles — absorption, centrifugal separation, electrostatically selective membranes, and coalescing, among others. The only thing that they all have in common is that they don't work particularly well on vegetable oil. And I include in this statement all the water blocking/separating filters that are currently sold by companies in the vegetable oil-as-fuel field.

This does not mean that they will not work at all — they will all work somewhat, but there is not one single filter designed to handle petrodiesel that I know of that has been shown to work well with vegetable oil. This is not shocking news, as vegetable oil, especially waste vegetable oil, is quite different from petrodiesel in the relevant fuel properties for these kinds of filters: viscosity, specific gravity, presence of emulsifiers, and polarity.

If anybody tells you that they have a filter where you put wet oil in one side and you get dry oil out the other side, ask them for the data. And if you are convinced, send the data to me, in care of New Society Publishers.

You don't need to remove water from the oil.

This actually covers two submyths. The first is that there is no water in oil straight from a fryer, so there's no water to remove. The second is that water is in the oil but it is not a problem. Let's have a look at each:

There is no water in the oil straight out of a fryer.

If you ignore reality, the reasoning behind this is actually pretty sound. It goes like this: Water boils at 212°F but a deep fat fryer operates at over 300°F, so obviously there's not going to be any water in the oil because any moisture would immediately boil away. However, this is not what we find. We find, and you can too with the hot pan test, that oil we have collected straight out of the fryer and let cool and sit for a couple of days has significant amounts of water in it.

What's probably going on is a little bit of two things. First, as the oil is used, acids, salts, and sugars contaminate the oil and change the boiling point of water so that it does not immediately boil away. Second, a certain amount of water is absorbed out of the atmosphere once the oil is left to cool, especially if the oil does contain a lot of acids, salts, sugars, and other similar compounds that are called hydrophilic, which just means water-loving. I'm not convinced that this is the correct understanding of why there is substantial amount of water in frying oil, but that doesn't change the fact that there usually is.

The amount of water in oil right out of a fryer won't cause a problem.

This is a slightly more sophisticated myth. Basically the idea is that if water didn't boil out of the oil when it was at frying temperature, the water won't come out period, or if it does come out, it will be in such small quantities that it won't create a problem. Again, this is a sensible theory, but an incorrect one. By the time you actually get around to processing the oil, you typically will be able to remove a significant amount of water just by letting the oil settle. If there is enough water in the oil for the water to appear just by settling, it certainly is not safe to use that oil in your vehicle without dewatering.

Settling and Filtering

Settling and filtering is still the most common way to remove water and particulate matter from waste vegetable oil. First, the oil is allowed to sit, either heated or unheated, to allow for the water to settle to the bottom of the oil. Oil is drained from the bottom and tested until it has an acceptable level of water, and is then passed through either a bag filter or cartridge filter.

Heating

Heating the oil to a moderate temperature, say around 90°F, helps break the emulsions of water and oil and dramatically shortens the time required to separate water and oil from weeks to hours. However, when heating the oil, care must be taken not to introduce convection currents that will mix the oil and prevent the water from separating. Vegetable oil is also heated in order to allow it to flow more easily through the filter.

When the oil is heated for filtering, the oil should not exceed the coolest temperatures reached in your conversion system. Vegetable oil is a mixture of triglycerides with different melting points, and if you filter the oil at a high temperature, you may have problems with vegetable oil solidifying in the cooler parts of your conversion system.

Filters

Bag filters are much preferred over cartridge filters, because of lower operating costs. Bag filters have a higher initial cost because the filter housings tend to be more expensive, but the cost per filter is much lower compared to cartridge-style filters. Cartridge filters designed for water filtration are *not* acceptable, as they bypass and fail when used for vegetable oil.

Filter Ratings

Filters are rated by the size of the particles they will filter, and whether the filter is absolute or nominal. A bag filter that is rated at 5 microns absolute will remove 90 percent of the particles larger than five microns. A filter that is rated to 5 microns nominal will filter particles but not to any specific degree of efficiency. In order for the bag filter to operate at its rating, it must be used as designed.

Abuse of Filters

Bag filters are designed to be supported by a cage in a filter housing. Using filters without a cage is a common and unwise practice. If the bag filter is not supported it will stretch and allow particles through much larger than the size the filter was rated for.

Filter Life

As a filter is used, trapped particles form a filter bed, which actually improves the efficiency of the filter in removing particles. Eventually, the filter bed will become so thick that fluid can no longer flow through the filter, and it will need to be replaced. Unless indicated by the manufacturer, bag filters are not reusable.

Do-It-Yourself Filters

In order to reduce costs, some people make filters themselves, particularly out of old jeans. They get jeans from a charity shop, cut the pant legs and sew the bottom of the leg closed. These homemade filters can certainly be useful as primary filter, especially since they can be washed and re-used; however they are inappropriate as a final filter since there is no way to gauge what size particles they are filtering out and at what efficiency.

Plans

In my opinion, the Frybrid Still is one of the best plans for a settling and filtering setup, in terms of cost and effectiveness. If you can get the hot water heater for free, the total system cost will be around $500.

This system allows you to fill a water heater unit with oil, heat it for several hours, then allow it to cool. This will precipitate the water and solids to the bottom where they may be drained off. The oil is then heated again and cycled through the filters for an hour or so, "polishing" the oil. The hot oil is then pumped from the heater unit into storage tanks or directly into the vehicle.

Figure 4.1 shows the basic plumbing.

The water heater will have four threaded fittings on it: H = Hot, V = Vent, C = Cold and D = Drain.

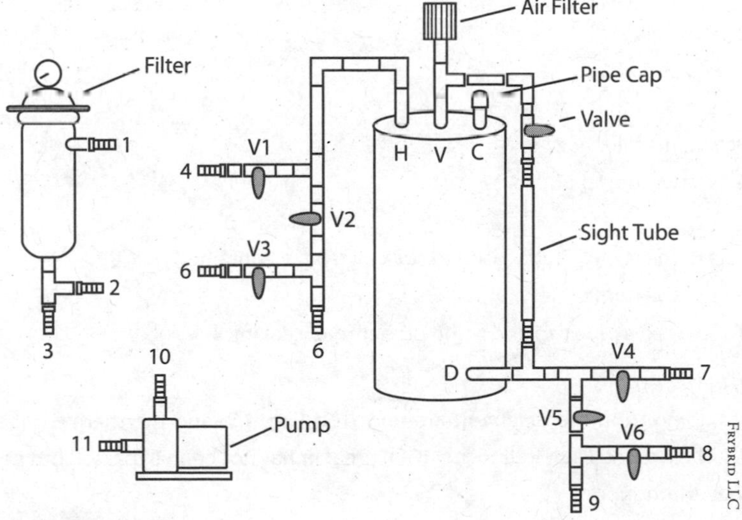

Fig. 4.1: *The Frybrid Still.*

You will need the following components:

Item	Quantity	Suggested Source
80 gallon hot water heater	1	Salvage
¾″ hose – ¾″ NPT hose barbs	13	Plumbing supply
¾″ NPT Street "Tees"	6	Plumbing supply
¾″ NPT Ball Valves	7	Plumbing supply
¾″ Poly Tube (clear with braiding)	6′	Plumbing supply
¾″ Black Iron Pipe		Plumbing supply
¾″ Close Nipples		Plumbing supply
Bag Filter Housing	1	McMaster –Carr #6870K57
Self Priming Pump	1	Northern Tools #36057
Fill nozzle	1	Strick Equipment or AW D

FRYBRID LLC

Table 4.2: *Frybrid Still Components.*

Water heater: I use an 80-gallon tank because it is what I found on the roadside. It needs to be a 220VAC model, but wired to use 110VAC so that the elements run cooler (oil needs one-third as much energy to heat as water does). Disconnect the upper element so that it does not function (you don't want the element getting hot when the oil level is below the element). *Never* turn on the heater when the oil level is below the lower element. Set the thermostat to 140°F.
Prepare yourself for a dozen trips to the hardware store.

Plumbing is as follows:

1 connects to 5
2 connects to 4
3 connects to the filler nozzle
6 connects to the pump outlet
7 is the drain
8 connects to the suction tube (used to suck oil from a container)
9 connects to the pump inlet
You will also need a pipe cap and some pipe thread sealant.

Rewiring the heater:

The wires going to the top element are removed (Fig. 4.2) and the thermostat is bypassed. The Ground Fault Interrupter (GFI) breaker has not been bypassed but has been wired through. ☞

Graydon Blair

Fig. 4.2: *The top element, with wires removed.*

Graydon Blair

Fig. 4.3: *The lower element, with wires and thermostat connected.*

The bottom element keeps its thermostat circuit (Fig. 4.3), and the internal wiring is connected to the power supply at the top well with wire nuts (Fig. 4.4). The ground circuit (green wire) is fastened through the ground lug on the heater.

Notice: Frybrid has provided these images and text as an example of what we have done. Frybrid in no way accepts any liability arising from the use of these concepts. The risk of fire, electric shock, or damage to persons or property rests solely on the person building such a device. Remember this design uses 110VAC to heat up to 80 gallons of oil to temperatures which can burn flesh. Anyone who has ever seen a grease fire will attest to how fierce they can be; you only have to imagine an 80-gallon grease fire. Never plug the heater in unless the level of oil in the heater fully covers the heating element. Never leave this unit unattended when plugged in. ☞

Graydon Blair

Fig. 4.4: *Power supply is 110-volt AC, connected to internal wiring with wire nuts. The system must be properly grounded.*

Function is as follows: Oil should be settled and as clean as possible. I have a pickup tube with a screen on the end. The pickup hose then connects to a Marine Raw Water Strainer before it runs through the pump and into my transport container.

Fill: Valves V6 and V2 are open, all others are closed. Suction tube is placed into container of oil and pump turned on. Oil is sucked up through the suction tube and pumped into the "still". When you have all the oil you need, pull the suction tube out of the container with the pump running and listen for a change is the sound of the pump. This will clear oil from the suction tube so it doesn't drain out on your floor.

Heat: Close all valves and turn the heater on after making sure that the heater is at least half full. Wait about two hours. The oil should have heated in the first hour to about 140°F, and after this point the heater's thermostat should maintain that temperature.

Cool and Separate: Turn the heater off and let it sit overnight. During this time the heated oil will cool and heavy particulates and mush of the suspended water should drop to the bottom. Now open valve V4 (all others closed) and drain off the sludge and water at the bottom, when the oil looks like oil, shut the valve.

Filter: Turn the heater back on and wait about an hour for the oil to heat up. Once it's hot, open valves V5, V3, and V1 (all others closed) and turn on the pump. This will draw oil from the bottom of the heater, through the pump and push it through the filter and back into the heater. Allow it to cycle for about an hour.

Fill'er Up: With the oil hot and the heater turned *off*, open valves V5 and V3 (all others closed). Turn on the pump, place the fill nozzle in a container or vegetable oil fuel tank and open the handle on the nozzle. The pump will draw oil from the heater bottom, and pump it through the filter and out the nozzle.

Collecting Oil: For collecting oil, we like Tuthill's 12/24V DC Fillrite pump (FR1604). It can be obtained from Strick Equipment (strickequipment.co/catalog/elecoilpump.htm) or AW Direct (awdirect.com/finditem.cfm?itemid=20299).

Storage: Vegetable oil should be filtered to at least 5 microns (2 microns is preferred) and dewatered to 500ppm or less (checked using the crackle test). The oil should then be placed in a container with no air space in it and sealed. Oil to be stored for some time should be kept as cold as is easily possible; a basement or cellar should be adequate. Exposure to oxygen, sunlight, or heat will decrease the amount of time which vegetable oil can be stored. If stored oil has become quite clear and has a ☞

paint-like smell, discard it. When the oil is to be used, it should be pumped from the storage container, through a filter, and into the vehicle. Leave a small amount in the bottom of the container and never draw fuel from the bottom. This oil can be poured into the next batch of oil being prepared and reused.

All storage vessels should be thoroughly cleaned after filling and stored in a secure area to prevent the attraction of pests and the possible rupture of the storage containers by pests. The smell of vegetable oil will attract insects, rodents, and large carnivores such as coyote or bear. Keeping the stored oil containers and area clean is essential. ■

Frybrid: http:www.frybrid.com/still.htm

Advantages

The settling and filtering method is still the standard method for processing oil, and has been used to great success by hundreds of people. As a batch unit, it can process oil more quickly than centrifuge methods.

Disadvantages

This method has the continuous cost of filters. For this method, processing time does not depend very much on the amount of oil in a batch. It will take about the same amount of time to process 50 gallons as it takes to filter 80 gallons. When filtering less than 50 gallons at a time, centrifuge methods have a clear advantage in rate of processing.

Off-the-Shelf Options

There is currently no off-the-shelf product that I would recommend. Fossil Free Fuels is working on an automated unit that they hope to bring to market by the time this book is published. The basic 60-gallon unit is expected to sell for around $1,500. I have not personally evaluated this unit, but from what I've seen it is based upon sound principles, and Fossil Free Fuels has indicated they will be able to provide data on the efficacy of their product.

Centrifuges

Centrifuges have been discussed for a long time as a possible means of cleaning waste cooking oil that would be faster, possibly more effective, and less expensive than traditional cleaning setups, because they do not require disposable filters, But it is only recently that centrifuges have been demonstrated to be practical, effective, and relatively inexpensive means of cleaning waste cooking oil, due mainly to the work of Sunwizard.[3]

Below I will describe the system that Sunwizard built, and show some of the variations that others have investigated. I do not have personal experience with these kinds of systems, but I am convinced that they work.

The heart of the system is the centrifuge (Fig. 4.5), specifically the Dieselcraft OC-20. This centrifuge was originally designed to clean

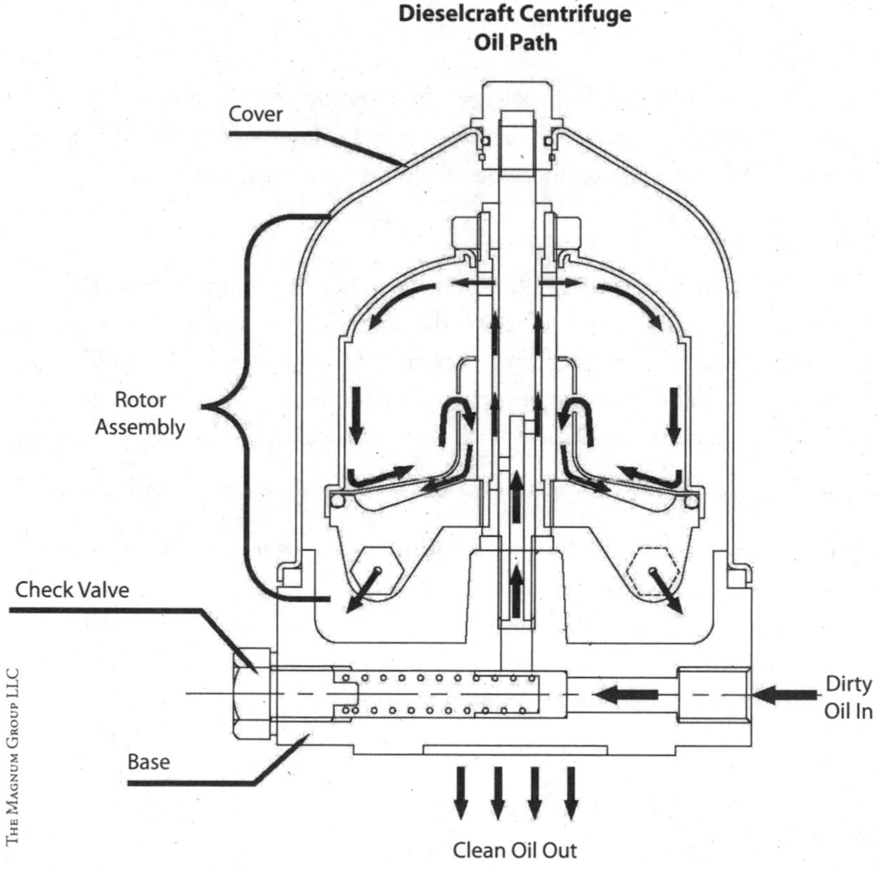

Fig. 4.5: *The centrifuge internal diagram.*

lubricating motor oil. Oil pressure spins the rotor at a rate over 6,000 rpm, creating centrifugal force that separates the solid contaminants and water from the main oil supply by forcing the contaminants into an outer bowl, which is easily cleaned. Water is also removed, to some degree, through evaporation. As part of the action of the centrifuge the oil is sprayed through jets which radically increases the surface area of the oil and promotes evaporation, especially as the oil is heated.

This particular centrifuge works best with a pump that can deliver 90 psi and a flow rate of .93 gallons per minute, or 55 gallons per hour, at this

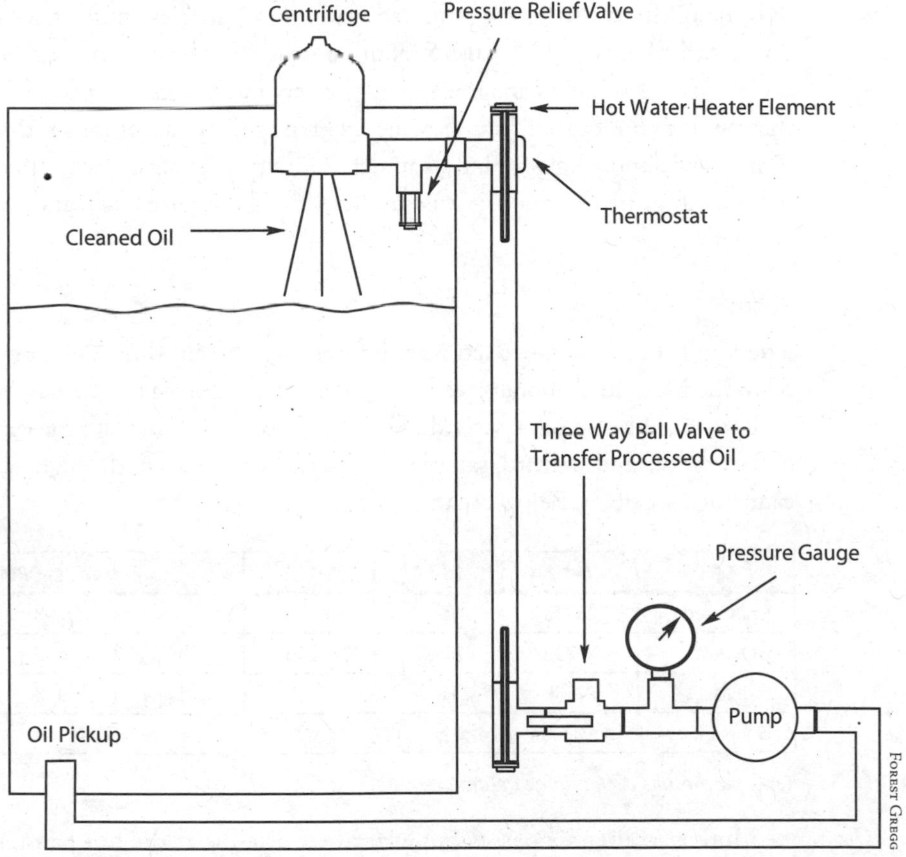

Fig. 4.6: *The centrifuge oil-cleaning setup, with pump.*

flow rate, the centrifuge should spin at about 7,000 rpm. Sunwizard originally used a vane-style power steering pump in conjunction with a ¾-horsepower electric motor, but later discovered that gear type pumps proved more durable than vane pumps, and motors providing as little as ¼ horsepower would work. He is now using a Oberdorfer 991N gear pump, that comes coupled to a 110-volt motor.

The third ingredient is heat. Waste vegetable oil tends to be too thick for the centrifuge until it is heated to 120°F, and Sunwizard found that heating the oil up to 160°F continued to improve how well the centrifuge separated water and particles from the oil.

A couple more notes about Sunwizard's system: First, he took care to pull the oil from the lowest point, so water and particles couldn't settle down and hide at the bottom. Second, before the pump is turned off, there must be some means to capture the separated water and particles that will drain out of the slowing centrifuge. To accomplish this, Sunwizard simply held a pint jar under the output at shutdown. Once the centrifuge had stopped, it was simply cleaned out with a rag, and was ready for reuse.

Results

The results of Sunwizard's setup are impressive. He sent three samples of oil to the Herguth Laboratories, Inc. in Vallejo, California to be analyzed — one of unfiltered oil, a second of oil that had passed through the centrifuge once, and a third sample that had been passed through the centrifuge six times. Below are the results:

	6-10	10-14	14-25	>25	Water PPM
Unfiltered Oil	23483	18067	8559	132	708
One Pass	29230	14542	1560	58	545
Six Passes	23438	1560	129	5	478
[Source: xyzworks.com/centrifuge.htm]					

Table 4.3: *Results of centrifuge treatment.*

Moreover, after six passes Sunwizard reported that the hot pan test showed no bubbles, though the lab showed that it still contained water contamination at the level of 478ppm. With this setup, Sunwizard is getting over 10,000 miles out of his final fuel filter, which is rated at 10 microns.

As the data shows, the centrifuge does not seem to be as effective at reducing the smallest class of particles. In order to remove these smallest particles, a conventional filter is necessary.

Rate of Filtration

Sunwizard starts with four passes through the centrifuge, which he reports is sufficient to clean most oil. After four passes, he stops the pump, and cleans the rotor of the centrifuge. If it is dirty, (indicated by a thick layer built up in the rotor) he will run the oil through the unit for another two passes or so.

If the oil is very dirty, it can take many more passes to clean the oil. Heating and settling is probably necessary in conjunction with the centrifuge in order to clean the oil in a timely and energy efficient manner.

Benefits

The centrifuge method seems effective at cleaning oil to an acceptable level without settling or using disposable bag filters or cartridge filters. At an average time of four hours to clean a 50-gallon batch of oil, it takes less time to clean the oil than to settle it and pass it through a bag filter, at least for batches of oil 50 gallons or less.

Downsides

A centrifuge system has many more moving parts than a settle and bag filter setup, and currently takes a good deal more skill and scrounging to build this kind of system cheaply. Dieselcraft has started selling vegetable oil processing systems, either as kits or pre-assembled, that cost between $1,425 for the OPS-20 and $2,950 for the OPS-50x2, the price depending upon the centrifuge size and whether the components are pre-assembled. Currently, this product from Dieselcraft does not include means to heat the oil, which experience has shown is critical.

Variations

Spin Clean and Spinner II are other similar motor oil centrifuges available from dieselproducts.com. Both Dieselproducts and Dieselcraft make centrifuges larger than the OC-20, which have a larger flow rate and about the same centrifugal force, and should speed cleaning. These larger centrifuges are more expensive.

It may also be a good idea to build a partial cover for the drum, to reduce exposure of the hot oil to air, and to lessen the risk of oxidation. However, since this unit removes water, at least in part, through evaporation, the centrifuge should not be completely sealed.

A cone-bottom drum would ensure that all the water and heavy sediment would drain.

Parts List

Table 4.4 shows the parts list for Sunwizard's setup. Where Grainger is listed as a source, it refers to W.W. Grainger, Inc., an industrial supply company with over 600 branches in the US and Canada (grainger.com).

Testing Oil

After processing, oil should be tested to make sure that the water content is at an acceptable level. Other tests are desirable, and even critical, depending upon the original source of the oil.

Water Tests

As we discussed in the chapter on water in the fuel properties section, water content should be below 700 parts per million, or pass the hot pan test.

Unfortunately, the only accurate way to measure total water content in vegetable oil is with a Karl Fischer analysis, which requires special equipment that usually costs over $5,000. Most people forgo the expense, and use a crude but acceptable test called the crackle test borrowed from the hydraulics industry.

Hot Pan (Crackle) Test

Here it is: put a droplet of oil onto a pan that has been preheated to 300°F, and observe what happens. If the oil audibly crackles when it hits the pan, or if it produces many large bubbles that don't clear up within a few seconds, then we say that it has too much water in it. If there is no crackle, and only a few small bubbles are formed which clear away within a few seconds, then the oil has an acceptable water content.

Item	Quantity	Price	Source
Dieselcraft OC-20 Centrifuge	1	$239	dieselcraft.com
Power Steering Pump	1	$20	Auto salvage yard
1 ¼ Horsepower 110V Electric Motor	1	$20	Auto salvage yard
1-½" pulley wheel for electric motor	1	$5	Auto salvage yard
½" V-belt	1	$15	Auto supply
Pressure Relief Valve 25-250 psi	1	$5	Grainger, Part # 6D915
Hydraulic Hose	5 feet	$13	Auto salvage yard
¾" or larger, high temperature hose	8 feet	$9	Auto supply
1¼" steel pipe	24"	$5	Plumbing supply
4500KW 240-volt water heater element	1	$8	Plumbing supply
½" to 1¼" NPT Reducer	1	$5	Plumbing supply
Reducer to attach PS output to 1¼" Tee	1	$10	Plumbing supply
1¼" steel Tee	1	$4	Plumbing supply
½" steel tube	8"	$5	Auto supply
Plumbing fitting	1	$15	Plumbing supply
Pressure Gauge	1	$9	Grainger, Part # 5WZ34
¹⁄₁₆" thick by 1¼" ID metal washer	1	$1	Auto supply
1½" ID washer with ⅛" x 1¼" O-Ring	1	$2	Auto supply
55 Gallon Drum	1	$20	Salvage
Approximate total cost		$410	
[Source: xyzworks.com/centrifuge.htm, and personal communication with John Nightingale of Dieselcraft, and Sunwizard]			

Table 4.4: *Sunwizard's centrifuge parts list.*

Obviously, this is a fairly imprecise and subjective method, but it's the best affordable procedure we have right now. Some people prefer a different temperature, but I don't think it matters much as long as it's above 300°F and you stick with whatever temperature you choose. Laser temperature sensors have become relatively inexpensive, and are very helpful in keeping your test temperatures consistent.

Calcium Hydride Test

Also becoming more popular are calcium hydride test kits. They are less subjective than crackle test, and give a decent ball-park number. They can be had for about $200 from Sandy Brae Laboratories.[5]

These tests work by calcium hydride reacting with water in the oil to form hydrogen gas which will increase the pressure in the sealed unit. This increase in pressure can be read as directly on the included pressure gauge as parts per million water.

There have been some quality control problems with this product, specifically that one of the reactants had significant water contamination present. This in turn will give an inaccurately high indication of water contamination in the actual testing. When using this test, it should first be run with solvent replacing the portion of vegetable oil, in order to determine the base level of water in the reactants. Running the test on the solvent and reactants themselves will determine a base moisture level. Care must also be taken to ensure that the vessel is completely dry.

Karl Fischer Tests

The gold standard for determining water content is the Karl Fischer test. The cost of the equipment to run the test is prohibitively expensive for most people, but there are labs that will run the tests for a fee.

Comparison of Test Methods

Sunwizard and Joe Beatty have done testing comparing the hot pan test and Karl Fischer test, and the Sandy Brae and Karl Fischer tests, respectively.

Sunwizard reported the following broad associations between the hot pan test and the Karl Fischer test. His results are in line with description of the test from the hydraulics industry.[6]

Karl Fischer, ppm	Hot Pan Test
0-500	No bubbles
500-2000	Bubbles present, increasing in size with higher water content
>2000	Visible liquid water in the oil
[Source: noria.com/learning_center/category_article.asp?articleid=301]	

Table 4.5: *Comparison of Hot Pan test and Karl Fischer test.*

Beatty found the following correlations between the Sandy Brae calcium hydride test and the Karl Fischer test. He found that the Sandy Brae unit indicated that the reactants had about 55 ppm water. The table below compares the results of the Karl Fischer test and the results from the Sandy Brae test minus the water measured in the reactants.

	Sandy Brae, water ppm		Karl Fischer, water ppm	
	Test 1	Test 2	Test 1	Test 2
Sample 1	205	215	168	179
Sample 2	355	355	432	420
Sample 3	465	485	474	481
[Source: Joe Beatty, frybrid.com/forum/showpost.php?p=60476&postcount=20]				

Table 4.6: *Comparison of Karl Fischer test and Sandy Brae test, adjusted for water content.*

Other Tests

If you are using oil that was used to cook meat, smells offensive, or displays other signs of advanced degradation, it is likely prudent to test free fatty acids of the oil. Free fatty acids above 3 percent are cause for concern.

While it would be very desirable to test for particulate contamination and oxidative stability, and at the least very interesting to measure flash point, iodine value, and viscosity, these tests are generally too expensive to justify for a small-scale operation. However, there are labs that will provide these tests, for a fee, and it may be worth the cost to get a sense of how well you are processing the oil and develop a better understanding of exactly what you are working with.

Real-Life Test

Although it is not a very encouraging thought, you are doing a great deal of testing of the quality of your oil whenever you use it as a fuel. If you are really doing an excellent job processing your oil, it should only be necessary to change your vegetable oil filter in your automobile as part of its regularly scheduled service. In an unscientific poll that we conducted, out of 42 respondents, only nine people were getting more than 10,000 miles out of their vehicle's vegetable oil filter, more than half were getting less than 5,000 miles. This is not good. The final fuel filter in your vehicle is just like any other filter, it will not catch every single particle that it is rated for. While it may make short-term economic sense to buy cheap automotive filters instead of possibly spending a lot more time and effort upgrading your fuel processing, it is asking for trouble in the long term.

Vegetable Oil and the Law

First of all, I am not a lawyer, and the following is not legal advice. If you are worried about the legal risks of using vegetable oil fuel, then consult an attorney. That said, I will try to describe my perception of the legal landscape of vegetable oil fuel in the United States, as of this writing. The legal landscape may change, and indeed I hope it does. I am completely ignorant of legal issues surrounding vegetable oil in other countries.

In Short

Let me start by saying that most people have not run into any legal problems from fueling with vegetable oil. They are quiet little mice who are careful not to attract the attention of the hawk. However, this is the deal, as I understand it:

First, it is illegal to sell vegetable oil as fuel in the US, without an exemption from the Environmental Protection Agency (EPA).

Second, it may not be illegal to burn vegetable oil fuel on public, federally funded roads, if the vegetable oil fuel was never bought and sold as fuel, though if the fuel is used in a business vehicle it probably is illegal.

Third, it is illegal to modify a vehicle to burn vegetable oil fuel if those modifications are used on a public, federally funded road, without a certification of conformity or an exemption from the E.P.A.

Fourth, you owe federal and state road tax for the use of publicly maintained roads, regardless of the fuel used. Many states have a mechanism for figuring out the road tax for fuels not taxed at the pump.

Fifth, the dumpsters behind restaurants belong to rendering companies. Once oil is put into those dumpsters, it becomes the property of the rendering companies. Restaurants are not in the legal position to give you permission to remove oil from these dumpsters, and doing so without permission from the rendering companies is theft. Moreover, depending upon the contract the restaurant has with the rendering company, a restaurant may not be legally allowed to give you their waste oil. Before you start picking up waste oil from a restaurant, have the restaurant cancel its existing contract with the rendering company.

Sixth, renders in some states have had legislation passed that requires anybody who picks up waste oil to have a rendering license.

Federal Law

The 1995 Clean Air Act tasks the US Environmental Protection Agency with controlling what kinds of after-market alterations may be made to a vehicle that travels on federally funded roads, and controlling what fuels can enter commerce.

Aftermarket Alterations

Generally, it is illegal to make after-market alterations to emission-related components of an automobile — this is considered illegal tampering. It is illegal to make these kinds of alterations yourself, and it is illegal for a company to sell components that are intended to be used in such an alteration.[7]

The only time that such alterations are not illegal is if they are part of an EPA-certified "clean fuel" conversion.[8] Unfortunately, at this time, no company can offer an EPA-certified "clean fuel" conversion, although at least two companies are working towards that goal.

Certification

I'm going to spend some time talking about the process of certifying a conversion kit as a "clean fuel" conversion, because it is probable that the rules and regulations around this process will have an enormous impact on the development of the field in this country.

Basically, as it stands now, the certification process that a vendor of a vegetable oil conversion system must undergo is prohibitively expensive for our small industry. For vehicles made before 1993, the company must show that the vehicle, burning either fuel (petrodiesel or vegetable oil) meets Tier 0 standards for emissions. For 1993 and later model years, the company must show that the vehicle meets the emission standards applicable when the vehicle was originally manufactured. There are very few labs that can perform all the tests and which are acceptable to the EPA, and these labs have waiting lists of more than a year and are very expensive to contract with.[9]

For vehicles made after 1996, the company must show that the On Board Diagnostics system is completely functional on both fuels. This requires that every fault condition be tripped. This is also very expensive.[10]

A contact in one of the few remaining liquid propane/compressed natural gas conversion companies has estimated that a typical certificate of compliance will cost around $100,000 when all is said and done. For a vehicle made before 1996, without On Board Diagnostics, a company might be able to get away with spending a mere $40,000.

Once you have gone through all the expense and time, you receive a certificate of conformity for one test group. This certificate allows you legally sell and install an alternative fuel conversion system for all the vehicles in the test group. Unfortunately, a single model made in a specific year can be made up of over ten test groups, based upon differences in engine size, drive train, and gross vehicle weight. So for instance, a company gets a certificate of conformity for a 1998 Dodge Ram 2500 extended cab, long bed, with 4-wheel drive. That means that company can sell kits for a 1998 Dodge Ram 2500 extended cab, long bed, with 4-wheel drive, but probably can't sell kits if the truck has a short bed, or a regular cab, or two wheel drive. All those configurations require a separate certificate of conformity.

The EPA's policies on certificates of conformity were developed to deal with the Liquid Propane and Liquid Natural Gas conversion industry, and it is instructive to look at how EPA regulation shaped that field.

LPG and CNG Conversions Industry: A Case Study

Back in the 1970's, during the last energy crisis, LPG and CNG conversions were promoted as a good alternative fuel source that reduced emissions and burned a fuel of primarily domestic origin. An industry sprung up that in many ways resembles the vegetable oil fuel field today, many small kit makers spread throughout the country selling kits of various qualities intended for installation by the end consumer.

In the 1990s, the EPA started to tighten up the process of acquiring a certificate of conformity, and by early 2000, the industry had completely changed. In this country, it is now impossible to buy a kit to convert an on-the-road vehicle to burn LPG or CNG, though there are a handful of companies that sell conversion kits for explicitly off-road use. There are also a few large companies that have commercial fleets as customers, which retrofit the vehicles at their own facilities. These larger companies have held on because there are significant tax credits for companies that are using a certified "Clean Fuel," a distinction that, as we'll discuss, vegetable oil does not enjoy, and is unlikely to ever receive.

Certified Fuels

The EPA is also tasked with controlling what fuel can enter commerce. Basically, in order for a fuel to be legally sold it must be certified that vehicles burning this new fuel meet emissions standards. The testing necessary to receive this certification is probably on the order of a few million dollars. Further testing is required to certify a fuel as a "Clean Fuel," which brings with it a number of very attractive tax breaks to end users. It is unclear whether a fuel that has not entered commerce, and has never been bought or sold as fuel, is under the jurisdiction of the EPA. Right now, there is no entity or organization with the incentive and the money to undergo the process of having vegetable oil certified as a fuel that may be sold in commerce.

However, it is possible to obtain an exemption from the EPA in order to sell vegetable oil as fuel. To my knowledge, Albuquerque Alternative Energies is the only company that has gone through this bureaucratic nightmare.

Regulation and Enforcement

There is always a difference between regulations and the enforcement of those regulations, and the picture that I have sketched out is much worse than what is actually happening on the ground. The EPA is certainly aware of vegetable oil conversions, but they seem to, at present, taken the position that this is a young field and needs time to develop before enforcing regulation. In addition, the EPA has given exemptions in at least one case to allow a company to sell vegetable oil as fuel, and as of this date has only required one vendor to begin the certification process.

Moreover, the EPA has never, to my knowledge, targeted individual consumers for enforcement of regulations on "illegal tampering." However, any company that is interested in using vegetable oil for fuel in their business is well advised to stay in close contact with their local EPA office, and to start a conversation early.

However, the regulatory environment is certainly discouraging investment into this field, so necessary for development. It is my profound hope that as the EPA and the vegetable oil community come to know one another that policies might be changed in order to promote the health of this alternative fuel option while still allowing the EPA to fulfill its crucial function of protecting the quality of our environment. Given the history, it does not make me terribly optimistic.

Internal Revenue Service

The IRS requires that you pay road taxes on vegetable oil fuel. As of this writing, the form that you need to in order to pay the taxes is form 720, "Quarterly Federal Excise Tax Return," and section 60c is the relevant field. There is no exemption.

State

States differ in how they are approaching vegetable oil fuel. In general, most states seem happy as long as you pay road tax. Contact your local tax collecting body for details of how to submit payment. There are a number of cases of folks publicizing using vegetable oil fuel and quickly receiving a visit from the local revenuer.

Both Alabama and California require an individual to have a renderer's license to pick up waste oil from restaurants.

Rendering Companies

The dumpsters behind restaurants belong to rendering companies, and you should assume that the oil in those dumpsters do as well. If you take oil from those dumpsters without the permission of the restaurant and the rendering company, that is theft. It's not completely settled whether oil in the dumpsters belongs to the restaurant, and whether the restaurant has the legal right for you to remove oil from a rendering company's dumpster. No matter the law, the rendering company will almost certainly interpret it as theft, and it's a practice that should be avoided.

Depending upon the contract, a restaurant may also not be legally allowed to give you the waste oil, or do anything with it except put it into the dumpster.

Rendering companies know about folks using vegetable oil fuel, and generally they are not happy about it, characterizing the community as a bunch of thieving, tax-evading, mess-making ne'er-do-wells. While that description may flatter us, it doesn't serve us well to antagonize this industry. They are more powerful than we are and have already been successful in having legislation passed in some states to make life harder for us.

When you approach a restaurant about picking up oil, have them cancel the contract with the renderer and have the renderer remove their dumpster. It may be possible for the restaurant to keep their contract with the rendering company and to also give you oil, but it is generally better in the long term to be the sole collector of oil

Co-ops

Cooperatives can save you time and money and put you in a better position to compete with biodiesel refineries and renderers for waste vegetable oil. In addition, cooperatives allow the collection and dispensing of large amount of straight vegetable oil intended for fuel without the legal difficulties of selling a fuel uncertified by the EPA.

For the first time, there are now enough people in the straight vegetable oil movement to allow for local pooling of resources for collection, processing, and dispensing of waste vegetable oil. And unlike just a few years ago, the success of biodiesel has meant new competition to the oil that restaurants throw away. In some places, waste vegetable oil is hard to find, and competition for the oil is increasing everywhere with the high prices of yellow grease. These trends will continue.

What is a Cooperative?

Basically, a cooperative is a collection of people who pool their time and money to accomplish something that would not be as economical to do by themselves. Usually, each member enjoys the benefit of the collective effort in proportion to how much they contributed. A cooperative can be as simple as three friends informally exchanging complementary abilities and skills. It can be six people pooling money to build and buy filters for a filtration system they all use to filter oil they individually collect. Or it can be a 40-plus member organization with written by-laws, dues, and timesheets. Our case studies describe organizations that run the gamut.

As it true of most enterprises, the amount of formal organization is related to how small the collection of people is and how well they know each other. It seems like good practice to start the co-op with the minimum number of people needed to accomplish your objective, and be very careful and selective about expanding. As you get larger, you may have to incur expenses that you had not anticipated, such as a pump that meters fuel so you know more accurately how much individuals are taking, have longer meetings to discuss the business of the cooperative, and use larger equipment and more involved safety procedures. On the other hand, being bigger can mean that labor can potentially be spread out more thinly, there is more money for better fuel processing and collection systems, and you have more leverage when approaching officials. Organizations larger than four or five seem to be subject to the 80/20 rule, where 20 percent of people are vital and do most of the work, and 80 percent contribute a great deal less.

Legal Concerns

As we've already noted, it is illegal to sell vegetable oil as fuel without an exemption from the EPA, which is difficult to acquire. You may be tempted to sell it anyway, without an exemption, but I would not recommend it, even if you disclaim that you are not selling it for use as a fuel for an on-the-road vehicle. The EPA has not come down on anybody yet, but State Revenue offices have.

This does not mean that money cannot pass hands within your co-op, or even, if you decided to, that someone couldn't get paid to run portions of the operation, but it does mean that you should not be able to simply exchange money for vegetable oil. If you have written by-laws, it should be clear that you are not selling vegetable oil.

Case Studies

Below are three case studies of different attempts to form co-ops.

Case 1: The Restaurant Owner, the Oil Collector, and the Mechanic

After Hurricane Katrina, two friends in a mountain state, one a mechanic and the other a restaurant owner, were talking about the high price of fuel and what they could do about getting out from underneath it.

The restaurant owner brought up that he had heard of using vegetable oil as a fuel to power a diesel engine, and he had quite a supply of it from his restaurants. The mechanic was pretty interested and he started doing research online and came across the Frybrid forums (frybrid.com/forum/) where he found someone in his city who was also interested in converting his vehicle and coincidentally had just gotten into the oil-collecting business and had bought a truck that pumps up waste oil. Pretty soon the relation was set up, each of the three contributing their skills and resources. The mechanic contributed his labor to converting and maintaining the vegetable oil system. The restaurant owner contributed his oil. The oil collector collected and filtered the oil through a Frybrid still. They are talking about expanding the co-op, but only to folks who could really contribute something to the mix.

Case 2: The Farmers

In 2004, a handful of farmers in northern California became interested in running vegetable oil in their tractors. They were fortunate in finding an excellent source of oil in a local bottling plant that had a large amount of drippings; basically unused oil that was commercially unusable because it was an uncontrolled mixture of various vegetable oils. The farmers formed the Biofuels Research Cooperative that eventually grew to over 40 members. The cooperative has written by-laws and holds regular meetings

Case 3: The Conversion Shop

From the early days of Frybrid, we ran a pretty informal co-op. A few friends who had converted their systems picked up oil and then we would filter it. We decided to expand the co-op somewhat and make it a little more formal. We have about six members, who now pay $20 dollars a month for operating expenses and sign in how much oil that had brought and how much oil they took. They were also expected to take away trash and to help process the oil. The co-op worked moderately well, but Frybrid still spent much more time cleaning up messes and processing oil than the other members. It is probably unavoidable that whoever hosts the processing system is going to be saddled with more work, and this should expected on the front end, and the co-op arranged accordingly.

Appendix A:
Eleven Burning Questions (With No Answers As Yet)

There is a great deal about vegetable oil as a diesel fuel that is not currently known. Below is a list of these unanswered questions.

1. How does the ignition delay of heated vegetable oil compare, quantitatively, to unheated petrodiesel?
2. Why does heating vegetable oil produce a narrower spray pattern with a longer penetration rate, when we would expect that the decrease in viscosity should produce a wider spray pattern and a shallower penetration rate?
3. How do the emissions of a range of engines fueled with properly heated vegetable oil compare to emissions of those engines fueled with petrodiesel?
4. Why does oil with very high free fatty acids tend to gum up in the fuel system?
5. Is 160°F truly the minimum acceptable temperature for injection?
6. Can yellow grease be processed to an acceptable fuel standard, economically?
7. What difference is there between one-tank and two-tank systems in the long-term health of a variety of engines?
8. Can we show that an "adequate" conversion does not have a negative effect on the long-term health of a variety of engines?

9. What advantages can be gained by changing injector design, injector pressure, and timing?
10. What level of oxidation is acceptable in fuel?
11. Is there an economical additive we can use to slow down oxidation of vegetable oil?

Appendix B: Fuel Standards

There have been a handful of attempts to develop a standard for vegetable oil fuel. None are completely adequate, but each is useful to some degree.

The SWRI Specification

In the early 1980s, the USDA awarded a grant to the Southwestern Research Institute to develop a fuel standard for vegetable oil. I have been unable to get a copy of the final report, but the significant findings of the work were summarizing in an article in the Journal of the American Oil Chemist's Society.[1] The strong points of this specification are (1) that the viscosity and linolenic specifications were clearly based upon engine studies and analysis of spray pattern, and (2) the specification sets limits on viscosity at the temperature that the researchers thought adequate for injection of vegetable oil fuel, instead of at the typical diesel viscosity test temperature of 104°F. The weak point in the specification is that for water content and flash point, the researchers just borrowed the limits from the existing petrodiesel standard, resulting in limits that may have no relation to the actual properties of vegetable oil.

Pre-Standard DIN V 51605

In 2002, as a graduate student, Edgar Remmele laid the groundwork for this standard, which is currently in the process of being accepted as a full standard by the German Standards Board DIN.[2] The specification, in its current pre-standard form, represents the work of a consortium of oil crop produces, processors, professors, and vegetable oil conversion manufacturers. Its strong point is its extensiveness and sensitivity to properties that are only relevant for vegetable oil fuel. Its main weakness is that it is only intended to set as standard for one vegetable oil, rapeseed oil. The maximum recommended Iodine Value excludes many oils common to the US, particularly soybean and sunflower, both of which can be excellent fuels. In addition, some of the limits seem weakly supported and more a function of what can be easily produced than what has been shown to be safe in engines. The limit of water is a case in point.

Pryde Specification

Everett Pryde was a key figure in American research on vegetable oil fuel during the 1980s, organizing a number of important conferences and conference sections, as well as being the lead author of many original contributions from his base at the USDA Northern Regional Research Center. He published a very early first step towards a fuel standard for vegetable oil in 1982.[3] The specification is probably the least well supported of the three, but his concern with phosphorus-containing gums and waxes are still probably useful.

ASTM Petrodiesel and Biodiesel

In the US, petrodiesel and biodiesel are supposed to meet standards controlled by the ASTM. The standards for the two fuels are D 975-07a and D 6751 – 07a, respectively.

Physical Properties		SWRI Specification	Pre-Standard	Pryde Specification	ASTM Petrodiesel	ASTM Biodiesel
Cetane Number, min		35	39	-	40	47
Heat of Combustion, min	MJ/kg	39	36	-	-	-
Viscosity @ 140°C, max	centistokes	5	-	-	-	-
Viscosity @ 40°C, max	centistokes	-	36	30-50	-	1.9-6.0

Physical Properties		SWRI Specification	Pre-Standard	Pryde Specification	ASTM Petrodiesel	ASTM Biodiesel
Specific Gravity min-max		-	0.9-0.93	0.91-0.93	-	-
Pour Point, max	°C	-	-	-5	-	-
Cloud Point, max	°C	22	-	20	-	-
Flash Point, min	°C	52	220	-	52	93
Copper Corrosion Rating, max		3	-	-	3	3
Particulates, max	mg/L	8	-	-	-	-
Ash, max	% weight	0.01	0.01	0.05	0.01	0.01
Insolubles, max	% weight			0.001	-	-
Carbon Residue, max	% weight	-	0.4	-	-	-
Contamination, max	mg/kg	-	24	-	-	-
Wax, max	mg/kg	-	-	20	-	-
Sulphur, max	mg/kg	-	10	-	50	50
Magnesium and Calcium, max	mg/kg	-	20	-	-	5
Phosphorus, max	mg/kg		12	20	-	1
Volatile matter, max	mg/kg	-	-	300	-	-
Acid Number, max		-	2	-	-	0.5
Free Fatty Acid, max	% volume	-	1	0.2	-	0.25
Iodine Number		<135	<125	80-145	-	-
Linolenic/ Linoleic ratio, max		0.07	-	-	-	-
Linolenic acid, max	% weight	5	-	-	-	-
Water, max	% volume	0.01	0.07	0.2	-	-
Water and Sediment, max	% volume	-	-	-	.05	.05
[Source: ASTM Standards D 975-07a and D 6751 – 07a]						

Table B.1: *Comparison of Vegetable Oil Fuel Standards.*

Appendix C:
Guide to Diesel Vehicles

Since the mid-1980s, American consumers who want diesel vehicles have had limited options. You could buy a passenger vehicle from Mercedes or Volkswagen, or a full-size truck, van, or SUV from Ford, General Motors, or Chrysler/Dodge. These offering make up the vast bulk of the existing stock of diesel vehicles in this country, and we'll go into some depth describing them.

From the late 1970s up to about 1986, during the last energy crisis, nearly every manufacturer was offering diesel engines in every class of vehicle, and if you search hard, you can still find diesels from this era. However, we will not describe most of the vehicles from this era in detail. For most people, they will not represent a good value. Regardless of their original quality, they tend to be expensive prospects, because parts are hard to get and very few people know how to work on them.

Passenger Vehicles

Mercedes-Benz

Mercedes diesel sedans, coupes, and station wagons made between 1968 and 1985 contain some of the most durable and vegetable oil-friendly engines ever made. After 1985, the engine design changed, durability suffered, and the OM603 series produced in the early 1990s had manufacturing defects. In 1995, Mercedes introduced the OM606 engine, which has proven its reliability, but is very difficult to convert properly.

Year	E-Class						S-Class	
'68	115	220D OM615.912						
69								
70								
71								
72								
73			240D OM616.916					
74								
75				300D OM617.910				
76								
77	123		240D OM616.912	300D OM617.912				
78					300CD OM617.912		116	300SD Turbodiesel OM617.950
79						300TD OM617.912		
80								
81							126	300SD Turbodiesel OM617.951
82				300D Turbodiesel OM617.952	300CD Turbodiesel OM617.952	300TD Turbodiesel OM617.952		
83								
84								
85								

Sources:
oilburners.net/forums/showthread.php?p=109014
home.hiwaay.net/~gbf/mbmodels.html

Table C.1: *Great Mercedes Diesels.*

The OM61x series of engines are legendary for durability, and to this day, there are probably more passenger diesels on the road in this country with this engine than any other, and they are excellent candidates for conversion, with indirect injection, oil-lubricated Bosch injection pumps, and voluminous trunks for vegetable oil fuel tanks. This engine series was available in Sedans (220D, 240D, 300D, 300SD), coupes (300CD), and station wagons (300TD).

After 1985, Mercedes introduced the OM60x series, which had little in common with the previous design. The big changes were an aluminum head instead of the OM61x's cast iron, and hydraulic lifters. It took

Forest Gregg

Fig. C.1: *The Mercedes 300SD sedan, an excellent candidate for vegetable oil conversion.*

Mercedes a while to work out the kinks in the new series, and the OM603.970 and OM603.971 produced between 1990-1994, in particular, had improperly cast and weak connecting rods.

Starting in 1995, with the OM606 series, Mercedes seems to have worked out most of the problems, and these engines have proven to be reliable. Unfortunately, these later engines are much more complicated and difficult to work on than early designs making them very poor candidates for conversion for most people.

None of the Mercedes engines mentioned require the addition of a lift pump.

Volkswagen

Volkswagen diesel engines started out simple, reliable, and fairly underpowered in the 1970s, and they have gotten increasingly better through the years. Modern Volkswagen TDIs are very fuel efficient, very quiet, very responsive, and very powerful.

All the engines Volkswagen has made, to date, seem to take to vegetable oil very well, however beginning with the TDIs of 1990s, these vehicles do become harder to convert simply because there is often very little room under the hood to install components. Unlike all earlier designs, the conversion of PD-type TDIs requires the addition of lift pump for

	Rabbit/Golf			Jetta			Dasher/ Quantum/ Passat		New Beetle	Vanagon	Caddy
'77	Rabbit/Golf I	1.5					Dasher				
78											
79								1.5			
80				Jetta I							1.5
81		1.6			1.6					1.6	1.6
82							Quantum				
83			1.6 turbo			1.6 turbo		1.6 turbo			
84											
85	Golf II			Jetta II							
86											
87											
88											
89											
90							Passat III				
91					1.6 ECO diesel						
92											
93	Golf III			Jetta III	1.9 TDI VE Type						
94							Passat IV				
95											
96								1.9 TDI VE Type			
97											
98							Passat V		1.9 TDI VE Type		
99	Golf IV	1.9 TDI VE Type		Jetta IV							
00											
01											
02											
03											
04		1.9 TDI PD Type			1.9 TDI PD Type			2.0 TDI PD Type	1.9 TDI, PD Type		
05				Jetta V			Passat VI				
06	Rabbit V										

Table C.2: *Volkswagen Timeline.*

Forest Gregg

Fig. C.2: *A second-generation mid-1980s VW Dasher, also sold as a Quantum in some markets.*

vegetable oil, because the stock fuel pump is located in the tank. The PD-type also has a fuel temperature sensor that must be considered.

Also, because modern Volkswagen diesels are already so fuel efficient, and slow to come to operating temperature, it may not make financial sense to convert them unless your driving is characterized by long trips.

American Trucks, SUVs, and Vans

Since the 1980s, Ford, Chrysler/Dodge, and General Motors have offered diesel options for their full-sized pickup and van lines, as well as some SUVs through the years.

In general, Ford and Dodge have provided very durable and powerful engines, well suited for vegetable oil up through the late 1990s, when the manufacturers redesigned the engines to meet tougher emissions regulations. Newer designs have not had problems with vegetable oil, *per se*, but have generally been less durable than the earlier generation. Designs continue to be changed rapidly.

The diesels offered by General Motors have never had the reputation of power and durability enjoyed by Ford and Dodge, and the designs have included a number of significant weak points throughout nearly every generation. However, if these are taken into consideration, these vehicles can represent a very good value as they are usually far cheaper than a Ford or Dodge and parts are inexpensive and plentiful.

Dodge[1]

Since the 1989 model year, Dodge has offered diesel engines from Cummins as options in the Ram 2500 and 3500 pickup trucks.

First Generation: 1989-1994.

This Cummins B-Series 5.9 liter unit was a direct injected, turbocharged, six-cylinder, inline engine. The first engines in Dodge applications came with Bosch VE Rotary Injection pumps, mechanical lift pumps, and no intercooler.

The B-Series 5.9 was originally developed for agriculture and commercial trucking, and had a maximum Gross Vehicle Weight Rating of 66,000 pounds (33 tons). The 5.9s fitted into the Ram trucks were significantly derated, but still had a great deal of power, and aftermarket alterations could easily increase the horsepower significantly. The stock transmissions have not always been able to handle the power of the engines, especially when the engine has been modified, and a significant transmission aftermarket has developed for the Dodge trucks. In mid-1991, an intercooler was added, increasing horsepower.

Second Generation: 1994-1998.

In 1994, Cummins dropped the rotary injection pump in favor of the Bosch P7100 Inline Injection pump, and increased the size of the radiator and intercooler.

This is widely considered the best engine ever sold in America in a pickup truck. It is extremely reliable, easy to service, and very easy to modify for increased power.

The main thing to watch for on these engines is the so-called "killer dowel pin." On rare occasions, a dowel pin from the upper timing case cover would come loose and if it fell into the timing gear, it could cause catastrophic damage. Later models fixed this problem.

Third Generation:1998-2003

The 5.9 was heavily redesigned. Among the many alterations were doubling the number of valves, from 2 per cylinder to 4, which is why an engine from this generation is often referred to as a 24-valve. And, rather unfortunately, a new electronically controlled rotary injection pump — the VP44 — was used.

The VP44 has a reputation as a temperamental thing. It does not particularly like aftermarket modifications and it is very sensitive to the pressure of the fuel supplied to it. Unfortunately, the electronic fuel pump of this generation is not near as durable as the mechanical one of early generations, and a great many VP44s have had to be replaced. This generation can burn vegetable oil well, but, in my experience, converting the vehicle often shows that the VP44 injection pump is on the way out. A more reliable "UPS" or "FedEx" injection pump (named for their service in these fleet trucks) can be had from the aftermarket.

It should be noted here that a fuel pressure indicator for fuel supplying the injection pump should be a required part of any vegetable oil conversion. Also, the addition of a robust electrical lift pump is strongly recommended for this engine, as the stock lift pump is not capable of producing the necessary fuel pressure when pumping thicker vegetable oil fuel.

Some 1999 and 2000 Rams have had problems with poorly-cast blocks cracking.

Fourth Generation: 2003-Present

In 2003, common rail technology came to the Dodge Rams. These engines are much quieter, more powerful, and are less polluting. They have also shown themselves to be reliable. Not many vegetable oil conversions have been done on this generation, but those that we have done have worked very well. These engine designs have the lift pump in the stock tank, so will require the addition of a lift pump for vegetable oil.

Ford[2]

Ford has offered diesel engines made by International/Navistar since 1982 in the F-Series 250 and 350 pickup trucks, E-Series Vans, and the Excursion SUV. All the Ford trucks can be converted without addition of electric lift pump, however the popular conversion style developed by Jason Crawford for 1999 and up Powerstrokes does use a second, powerful vegetable oil lift pump.[3]

First Generation:1982-1993

Ford contracted with Navistar, parent company of International Engine Group, to provide diesel engines for Ford F-Series trucks and E-Series vans. The first offering was 6.9 liter V-8, indirect injected, naturally aspirated engine with a Stanadyne injection pump.

In 1988, the engine was upgraded to 7.3 liters. In 1993, Ford offered an optional turbocharger. These engines are durable, and excellent candidates for vegetable oil conversion.

Second Generation: 1994-1998-½

Branded the Powerstroke, the 7.3-liter diesel offering had a fuel gallery that ran inside the head and unit electro-hydraulic injectors that delivered fuel directly into the cylinder.

Vegetable oil conversions can be a good bit simpler with these engines, because the fuel gallery in the head acts as a very effective final fuel heat exchanger.

Third Generation: 1998-1/2-2003

The main alteration in this version was deadheading the fuel gallery. In the earlier Powerstrokes, a certain amount of fuel from the fuel gallery was constantly being returned to the fuel tank. However, this heated diesel fuel led to problems with evaporative emissions, so Navistar changed the design so that no fuel exited from the gallery except through the injectors. There is a good deal of debate about what effect this has. In either case, these engines are very good candidates for conversion. In general the 7.3-liter Powerstrokes are highly regarded engines for their power and durability.

Fourth Generation: 2004-2007

In 2004, Ford introduced the 6.0-liter Powerstroke. The displacement was reduced for better fuel economy, the number of valves per cylinder were doubled, and aluminum alloys replaced much of the cast iron of the earlier generation. Unforutnately the new design has its problems. Ford has issued over 77 technical service bulletins — far above average — and there has been a great deal of tension between Ford and Navistar over liability for engine problems. It is not clear if that relationship will survive.

General Motors

Starting in 1982, General Motor's subsidiary, Detroit Diesel provided diesel engines for GMC and Chevrolet trucks and SUVs. The Detroit Diesel 6.2 and 6.5s did not have either the power or the reliability of contemporaneous offereings from Ford or Dodge. However, these engines were available in a wide variety of platforms, and were designed to be easily interchangeable with any other GMC V-8. They can be made reliable, and are the cheapest diesel engines in the used vehicle market. These indirect injection engines can be excellent candidates for vegetable oil conversion.

In 2001, General Motors replaced the 6.5 with a 6.6-liter diesel engine called the Duramax, made in partnership with Isuzu. The Duramax is a powerful but complicated engine, with common rail direct injection technology, 32 valves, and an electronic fuel injection control module that is cooled by fuel. The early Duramax had some problems with injector failure, and like all common rail engines, these engines are more sensitive to fuel cleanliness than earlier engines with lower pressure injection pumps. The Duramax has *not* been a popular candidate for conversion. In order to convert it correctly, diesel fuel must be continuously circulated through the fuel injection control module, even when the engine is fed by vegetable oil. That requirement adds significant additional complication and cost to the conversion.

All the GM diesel engines require the addition of a secondary lift pump, either because the stock lift pump is weak and prone to failure, as in the case of the 6.5s and 6.6s, or because of the necessity to keep diesel flowing through the electronic hardware for the Duramax series.

The Rarities

There have been a number of other diesel offerings to make it to our shores, some of which are very charming and have excellent engine designs. However, I would not recommend these vehicles to most people for the simple fact that they are very rare, and so parts are expensive if not impossible to find, and finding a mechanic that will work on these will be a difficult task.

Japanese and Domestic Compact Trucks

In the 1980s, American consumer enthusiasm for Japanese small trucks took U.S. automakers by surprise, and the automakers responded either by rebadging the imports, or by buying engines from their foreign competitors and dropping them into existing vehicle lines. So, all the small diesel trucks sold in this country during the 1980s, with the exception of the Volkswagen Caddy, had a Japanese engine in them.

Mitsubishi Mighty Max	1983-86	Also rebadged as Dodge Ram 50
Nissan 720	1983-86	
Isuzu Pu'p (1981-1987)	1981-87	Also rebadged as Chevrolet LUV (1981-82)
Chevrolet S-10	1983-85	Available with a Isuzu 2.2 diesel
Ford Ranger	1983-88	Available with a Mazda 2.2 diesel (1983-84), and a Mitsubishi 2.3 diesel (1985-88)
Mazda B2200	1982-85	
Toyota Pickups	1981-86	

Table C.3: *Japanese and Domestic Compact Trucks.*

Compact Cars

The story is similar with compact cars. The Detroit manufacturers had little experience with diesel engines in passenger vehicles, and even less know-how about small diesels in compact cars. So, again with the exception of Volkswagens, all the diesel compact cars were either Japanese or had Japanese engines.

Ford Escort/Mercury Lynx	1984-87	Available with a Mazda 2.2 diesel
Ford Tempo/Mercury Topaz	1984-86	Available with a Mazda 2.2 diesel
Nissan Sentra	1984-87	
Toyota Camry	1983-86	

Table C.4: *Compact Cars.*

Oldsmobile Diesels

Only one American auto manufacturer has every tried to build their own diesel engine for widespread use in passenger vehicles, and they only tried it once. Between 1978 and 1985, GM put Oldsmobile-produced V6s and V8 in nearly every passenger vehicle line, including Cadillacs, Malibus, Chevettes, and Impalas. These engines had a very high rate of failure of the head gaskets bolts which, in turn would often lead to damaged valves, pistons, connecting rods or crankshafts as coolant seeped into the cylinders. Whether the failure was due to defects inherent in the design, or general poor fuel quality combined with an inadequate fuel filter/water separator, these engines are widely credited with souring American consumers to diesel engines for a generation. That said, if the fuel cleanliness is ensured, these engines can go for hundreds of thousands of miles.

Miscellaneous

International Scout	1976-80	Nissan SD-33, turbocharged in 1980
Isuzu Trooper	1981-87	Same engine as Isuzu P'UP
Nissan Maxima	1981-83	
Audi 5000	1979-83	
Peugeot 504/505	1979-86	
Peugeot 604	1981-84	
BMW 524td	1985-86	
Lincoln Continental	1984-85	Used a non-turbo version of the BMW engine in the BMW 524
Volvo 200 Series	1979-85	Available with VW 2.0 or 2.5

Table C.5: *Miscellaneous Diesel Vehicles.*

Appendix D:
Fatty Acid Profiles of Common Vegetable Oils

	Palmitic 16:0	Stearic 18:0	Arachidic 20:0	Behenic 22:0	Lignoceric 24:0	Oleic 18:1	Ricinoleic*	Linoleic 18:2	Linolenic 18:3	Erucic 22:1	Iodine Value	Saponification Value
Peanut	11.4	2.4	1.3	2.5	1.2	48.3	-	32.0	0.9	-	80-106	200
Castor	1.1	3.1	-	-	-	4.9	89.6	1.3		-	82-88	203
Cottonseed	28.3	0.9	-	-	-	13.3	-	57.5	-	-	90-140	207
Crambe	2.1	0.7	2.1	0.8	1.1	18.9	-	9.0	6.9	58.5	?	?
Canola	3.5	0.9	-	-	-	64.4	-	22.3	8.2	-	?	?
Corn	11.7	1.9	0.2	-	-	25.2	-	60.6	0.5	-	10?-140	194
Sesame	13.1	3.9	-	-	-	52.8	-	30.[illegible]	-	-	104-120	210
Sunflower	6.1	3.3	-	-	-	16.9	-	73.7	-	-	110-143	192
Soybean	11.8	3.2	-	-	-	23.3	-	55.5	6.3	-	117-143	221
Safflower	8.6	1.9	-	-	-	11.6	-	77.9	-	-	126-152	190
Linseed	4.9	2.4	-	-	-	19.7	-	18.0	54.9	-	168-204	189

*Ricinoleic contains an additional alcohol group and is only naturally present in castor oil.

Source: Goering, C. E., A. W. Schwab, M. J. Daugherty, E. H. Pryde, and A. J. Heakin. 1982. Fuel Properties of Eleven Vegetable Oils. Transactions of the ASAE 25, no. 6: 1472-1483.

Table A: Fatty Acid Profiles of Common Vegetable Oils, Percentages by Weight.

Vegetable Oil Fatty Acids

Saturated Fatty Acids
(0 double bonds)

Myristic (14:0)

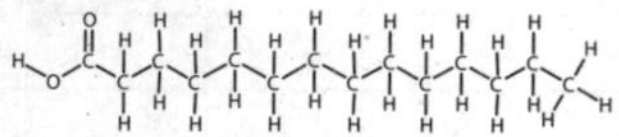

Palmitic (16:0)

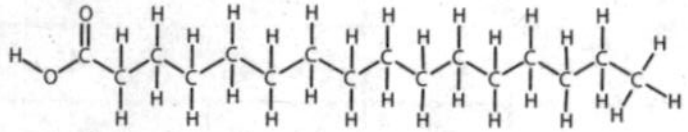

Stearic (18:0)

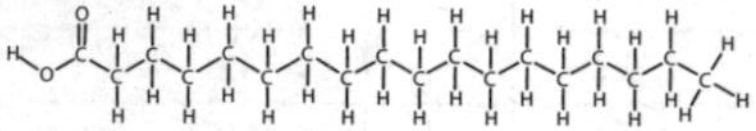

Arachidic (20:0)

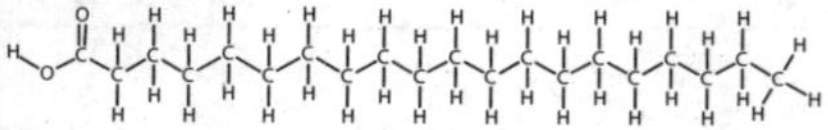

Behenic (22:0)

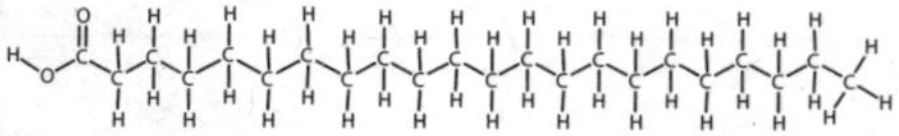

Lignoceric (24:0)

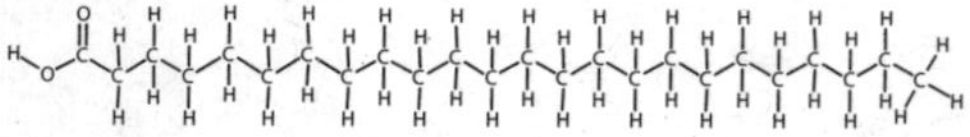

Monounsaturated Fatty Acids
(1 double bond)

Oleic (18:1)

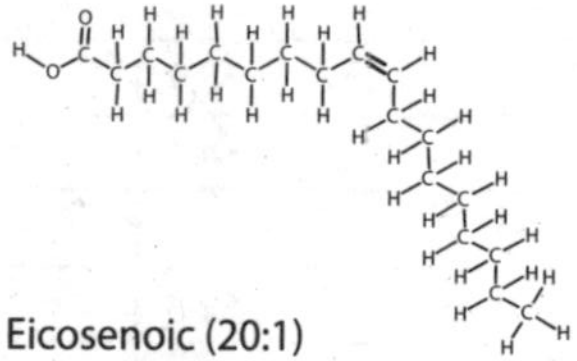

Eicosenoic (20:1)

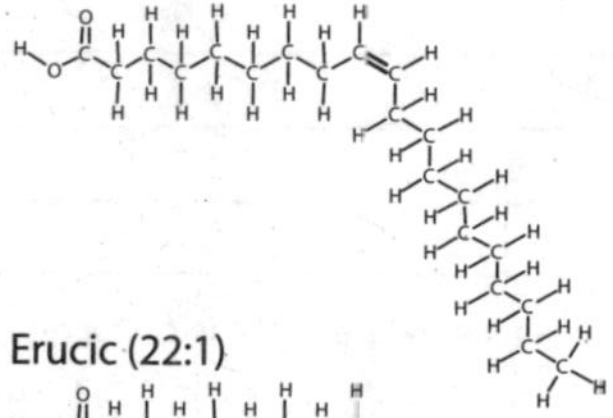

Erucic (22:1)

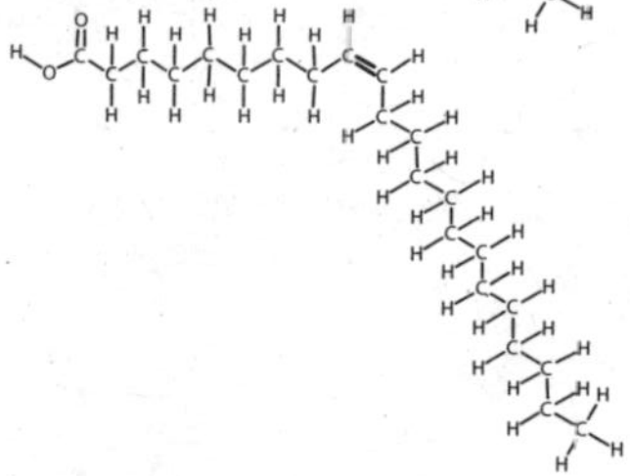

Polyunsaturated Fatty Acids
(2+ double bond)

Linoleic (18:2)

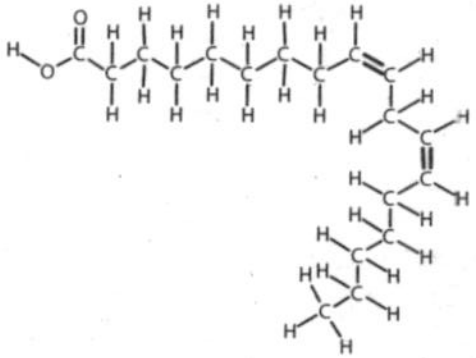

Linolenic (18:3)

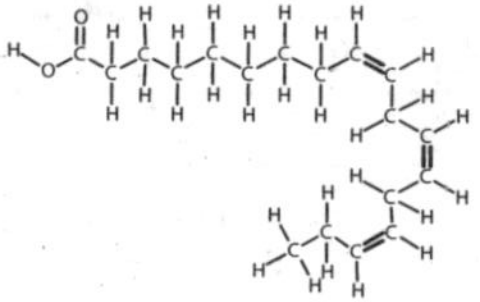

Ricinoleic (18:1)-OH

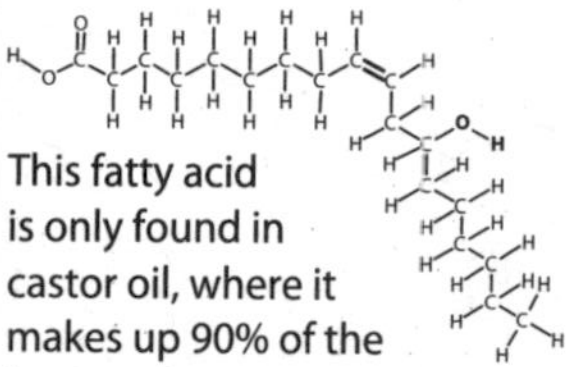

This fatty acid is only found in castor oil, where it makes up 90% of the fatty acid composition

The five bolded fatty acids — palmitic, stearic, oleic, linoleic, and linolenic — comprise over 95% of the fatty acid composition of most vegetable oils.

Annotated Bibliography

For the researcher, the four most important sources for this field are the *Journal of the American Oil Chemist's Society*, the *Transactions of the American Society of Agricultural Engineers* (ASAE), the *Society of Automotive Engineers (SAE) Technical Paper Series*, and the book: *Vegetable Oil Fuels: Proceedings of the International Conference on Plant and Vegetable Oils as Fuels, August, 1982*. St. Joseph, MI: American Society of Agricultural Engineers.

The first two sources are fairly accessible through most libraries. Of particular value are the August 1983 and October 1984 issues of the *Journal of the American Oil Chemist's Society*. These issues contain the proceedings of symposia dedicated to vegetable oil fuel that had taken place at the American Oil Chemist's Society annual conference.

Only universities and colleges with engineering departments are likely to have the *SAE Technical Paper Series* in their collection, and even they are almost sure to have significant gaps in the collection. Most SAE papers are available online from the SAE website, but at a very high cost.

Vegetable Oil Fuels should be available through interlibrary loan. Sometimes it is available for sale from online book merchants, but always at a high premium.

The bible for the commercial and industrial vegetable oil industry is *Bailey's Industrial Oil and Fat Products*, now in the sixth edition. You can buy the six-volume set new for $1,675, or piece together a collection of the previous editions for much cheaper. In any case, if you have a question about vegetable oil, it is likely that it is in *Bailey's*.

Below are the articles and books that I have found useful, interesting, or noteworthy, divided up by subject matter. Many articles could have been put into more than one category, but I chose to not duplicate the content, but indicate that fact in the description. I start out with three articles that I think I have found the most important.

Ryan, T.W., L.G. Dodge, and T.J. Callahan. 1984. The effects of vegetable oil properties on injection and combustion in two different diesel engines. *Journal of the American Oil Chemists' Society* 61, no. 10 (October 5): 1610-1619. dx.doi.org/10.1007/BF02541645

The U.S. Food and Drug Administration contracted with the Southwest Research Institute to develop a fuel standard for vegetable oil fuel. This paper presents most of the major findings of that research: including the surprising result that the spray pattern of vegetable oil gets narrower and penetrates deeper as it is heated, the effect of unsaturation on combustion, and a very solid recommendation for a fuel specification.

Baranescu, R.A., and J.J. Lusco. 1982. Performance Durability and Low Temperature Evaluation of Sunflower Oil as a Diesel Extender. In *Vegetable Oil Fuels: Proceedings of the International Conference on Plant and Vegetable Oils as Fuels*, 312-328. St. Joseph, MI: American Society of Agricultural Engineers.

This paper highlights the importance of bulk modulus in understanding the problem of carbon deposits arising from burning cold vegetable oil. This fact is critical and very little talked about in any other literature.

Stone, 1992. Combustion analysis of sunflower oil in a diesel engine and its impact on lubricating quality. *SAE Technical Paper Series:* 921631. aei-online.org/technical/papers/921631.

No real new results, but the best synthetic article of the field.

Review Articles

Bhattacharyya, S., and C. S. Reddy. 1994. Vegetable Oils as Fuels for Internal Combustion Engines: A Review. *Journal of Agricultural Engineering Research* 57, no. 3 (March): 157-166. dx.doi.org/10.1006/jaer.1994.1015.

Excellent review article. Separates research by oil type, discussing investigations done with safflower, rapeseed, sunflower, soybean, neem, Chinese tallow, palm, cottonseed, and peanut oils.

Gupta, M.K. 2005. Frying Oils. In *Bailey's Industrial Oil and Fat Products*, ed. F. Shahidi, 1-31. John Wiley & Sons. media.wiley.com/product_data/excerpt/92/04713854/0471385492.pdf. *Excellent overview of cooking oil from a food science perspective. Includes discussions of processing, reactions in the fryer, transportation and storage, and quality standards. Very good.*

Knothe, G. 2001. Historical perspectives on vegetable oil-based fuels. *INFORM - International News on Fats, Oils and Related Materials* 12, no. 11: 1103. *Excellent review of the history of vegetable oil fuels. Puts to bed the myth that Rudolph Diesel fueled his first engines on vegetable oil. Very good.*

Shay, G. E. 1993a. Diesel fuel from vegetable oils. Status and opportunities. *Biomass and Bioenergy* 4, no. 4: 227-242. dx.doi.org/10.1016/0961-9534(93)90080-N. *Reviews the potential for vegetable oil fuel, primarily for developing nations.*

Srivastava, A., and R. Prasad. 2000. Triglycerides-based diesel fuels. *Renewable and Sustainable Energy Reviews* 4, no. 2: 111-133. dx.doi.org/10.1016/S1364-0321(99)00013-1. *Good review of existing research, with focus on potential for vegetable oil as substitute fuels in India. The bulk of the paper is on biodiesel.*

Karaosmanoglu, F. 1999. Vegetable oil fuels: A review. *Energy Sources* 21, no. 3: 221-231. dx.doi.org/10.1080/00908319950014858. *A review of potential for vegetable oils as substitute fuels in Turkey and an overview of Turkish research in the field. Interesting as model for national analysis.*

Babu, A.K., and G. Devaradjane. 2003. Vegetable Oils and their Derivatives as Fuels for CI: An Overview. *SAE Technical Paper: 2003-01-0767*

Poor discussion of vegetable oil. Decent table of fuel properties for 19 vegetable oils.

Bandel, W., and W. Heinrich. 1982. Vegetable Oil Derived Fuels and Problems Related to their Use in Diesel Engines. In *International Conference on Biomass*, ed. A. Strub, P. Chartier, and G. Schleser, 822. New York, NY: Applied Science.

No deep insights, but still useful as brief but extremely lucid description of the promise and challenges of vegetable oil fuel.

Peterson, C. L., G. L. Wagner, and D. L. Auld. 1983. Vegetable Oil Substitutes for Diesel Fuel. *Transactions of the ASAE* 26, no. 2: 322-332.
Good early review of research by biofuel pioneer, ultimately pessimistic about use of straight vegetable oil as diesel fuel.

Peterson, C. L., D. L. Auld, and R. A. Korus. 1983. Winter rape oil fuel for diesel engines: recovery and utilization. *Journal of the American Oil Chemists' Society* 60, no. 8: 1579-1587.
dx.doi.org/10.1007/BF02666589.
Review of research on small scale production and use of rapeseed oil as alternative fuel. Pessimistic, good summary.

Fuel Properties

Goering, C. E., A. Schwab, M. Dougherty, M. Pryde, and A. Heakin. 1982. Fuel Properties of Eleven Vegetable Oils. *Transactions of the ASAE* 25, no. 6: 1472-1483.
Presents fatty acid composition, acid value, phosphorous content, peroxide value, viscosity, cetane number, higher heating value, cloud point, pour point, flash point, density, carbon residue, ash content, sulphur content, copper corrosion number, and induction period for petrodiesel and eleven vegetable oils: castor, corn, cottonseed, crambe, linseed, peanut, rapeseed, safflower, high oleic safflower, sesame, soybean, and sunflower. A number of equations are suggested for predicting viscosity, higher heating value, cetane, and induction period based upon average fatty acid chain length, average number of double bonds, and percentages of linolenic and linolenic acids. Very good information.

Demirbas, A. 1998. Fuel Properties and calculation of higher heating values of vegetable oil. *Fuel* 77, no. 9/10: 1117-1120.
dx.doi.org/10.1016/S0016-2361(97)00289-5.

———. A. 2003. Chemical and fuel properties of seventeen vegetable oils. *Energy Sources* 25, no. 7: 721-728.
dx.doi.org/10.1080/00908310390212426.
The first paper presents the distillation range, flash point, viscosity, carbon residue, cetane number, higher heating value, ash content, sulfur content, iodine value, saponification value, and fatty acid composition of twenty-two vegetable oils: cottonseed, soybean, rapeseed, safflower, sunflower, sesame, linseed, peanut,

wheat grain, corn marrow, castor, poppy, bay laurel leaf, hazelnut, walnut, almond, olive kernel, ailanthus, beech, beechnut, crambe, high oleic safflower, and spruce. The second paper presents the same data, though omitting the more obscure oils, and introducing errors into the properties table, it is not reliable. At the end of the first paper, Demirbas suggests an equation to predict higher heating value.

Coupland and McClements. 1997. Physical properties of liquid edible oils. *Journal of the American Oil Chemists' Society* 74, no. 12 (December 1): 1559-1564. dx.doi.org/10.1007/s11746-997-0077-1.

Reports the ultrasonic velocity, attenuation, specific heat at constant pressure, density, adiabatic expansion coefficient, dynamic viscosity, and thermal conductivity of nine vegetable oils at 20°C: sunflower, corn, olive, rape, cotton, peanut, palm, safflower, and soybean oil.

Canakci, M. 2007. The potential of restaurant waste lipids as biodiesel feedstocks. *Bioresource Technology* 98, no. 1 (January): 183-190. dx.doi.org/10.1016/j.biortech.2005.11.022.

Includes tables of American metropolitan waste oil resources and estimates of total waste oil, along with data on peroxide value, FFA, and contaminants for different waste streams of an Iowa rendering company over the period of one year.

Isigigur, A., F. Karaosmanoglu, and H. A. Aksoy. 1995. Characteristics of safflower seed oils of Turkish origin. *Journal of the American Oil Chemists' Society* 72, no. 10: 1223. dx.doi.org/10.1007/BF02540994.

Reports acid value, saponification value, iodine number, and fatty acid composition of Turkish safflower oils.

Goodrum, J. W. 1984. Fuel properties of peanut oil blends. *Transactions of the ASAE* 27, no. 5: 1257-1262.

Provides surface tension, density, and kinematic viscosity at 22°, 38°, and 54° for crude vegetable oil and blends with gasoline, methyl esters, and butanol ranging from 90 percent to 10 percent peanut oil. Also includes fitted coefficients for dynamic viscosity/temperature equations for all blends.

Karaosmanoglu, F., M. Tuter, E. Gollu, S. Yanmaz, and E. Altintig. 1999. Fuel Properties of Cottonseed Oil. *Energy Sources, Part A: Recovery, Utilization, and Environmental Effects* 21, no. 9 (September): 821-828. dx.doi.org/10.1080/00908319950014371

Provides density, flash point, pour point, cloud point, cold filter plugging point, sulfur content, ash content, higher heating value, and cetane for cottonseed oil.

Przybylski, R. Canola Oil: Physical and Chemical Properties. *Canola Council of Canada*. canola-council.org/Chemical1-6/Chemical1-6_1.html.
Detailed description of physical and chemical properties of canola oil.

Viscosity

Goodrum, J. W., D. P. Geller, and T. T. Adams. 2002. Rheological characterization of yellow grease and poultry fat. *Journal of the American Oil Chemists' Society* 79, no. 10 (October 23): 961-964. dx.doi.org/10.1007/s11746-002-0587-2.

———. 2003. Rheological characterization of animal fats and their mixtures with #2 fuel oil. *Biomass and Bioenergy* 24, no. 3 (March): 249-256. dx.doi.org/10.1016/S0961-9534(02)00136-8.
These two papers characterize the fatty acid composition and dynamic viscosity of beef tallow, choice white, poultry fat, and yellow grease.

Lang, W., S. Sokhansanj, and F. W. Sosulski. 1992a. Modelling the Temperature Dependence of Kinematic Viscosity for Refined Canola Oil. *Journal of the American Oil Chemist's Society* 69, no. 10: 1054.
Provides experimental and previously reported data on the kinematic viscosity of refined, bleached, and deodorized canola oil; refined, bleached, and winterized canola oil, and refined soybean oil at 5 different temperatures between 0° and 40°C. Also includes equation and fitted values for kinematic viscosity/temperature.

Santos, J. C. O., I. M. G. Santos, and A. G. Souza. 2005. Effect of heating and cooling on rheological parameters of edible vegetable oils. *Journal of Food Engineering* 67, no. 4: 401-405. dx.doi.org/10.1016/j.jfoodeng.2004.05.007.
Vegetable oil viscosity has a property called hysteresis, wherein a change in temperature does not immediately lead to a change in viscosity.

Ceriani, R., C.B. Goncalves, J. Rabelo, M. Caruso, A.C.C. Cunha, F.W. Cavaleri, et al. 2007. Group Contribution Model for Predicting Viscosity of Fatty Compounds. *Journal of Chemical & Engineering Data* 52, no. 3 (May 10): 965-972. dx.doi.org/10.1021/je600552b.
Develops equations to predict dynamic viscosity/temperature relationships from fatty acid composition.

Fasina, O.O., H. Hallman, M. Craig-Schmidt, and C. Clements. 2006. Predicting temperature-dependence viscosity of vegetable oils from fatty acid composition. *Journal of the American Oil Chemists' Society* 83, no. 10 (October 2): 899-903. dx.doi.org/10.1007/s11746-006-5044-8.

Presents experimentally determined dynamic viscosity/temperature relationships for twelve vegetable oils, fitted equations for those oils, and an equation for predicting the dynamic viscosity/temperature relationship based upon fatty acid composition. Very good work.

Halvorsen, Mammel and Clements. 1993. Density estimation for fatty acids and vegetable oils based on their fatty acid composition. *Journal of the American Oil Chemists' Society* 70, no. 9: 875-880. dx.doi.org/10.1007/BF02545346.

Presents equations that can be used to estimate density of vegetable oils at various temperatures. With this information, kinematic viscosity can be determined from dynamic viscosity and vice versa.

Viscosity of Typical Fluids vs. Temperature. 2003. In *The Lee Company Electro-Fluidic Systems Technical Handbook*, Release 7.1. The Lee Company Technical Center.

Besides the most thorough viscosity/temperature chart I have ever seen, this strange little catalog has a number of very interesting and somewhat useful descriptions of how to think about fluids moving through tubes and valves. The chart is also available online at tinyurl.com/2lqjdz.

Heating Values

Freedman, B., M. O. Bagby, and H. Khoury. 1989. Correlation of heats of combustion with empirical formulas for fatty alcohols. *Journal of the American Oil Chemists' Society* 66, no. 4: 595-596.

Freedman, B., and M. O. Bagby. 1989. Heats of combustion of fatty esters and triglycerides. *Journal of the American Oil Chemists' Society* 66, no. 11: 1601-1605. dx.doi.org/10.1007/BF02636185.

Presents higher heating values for fatty alcohols, methyl esters, ethyl esters, and triglycerides with both saturated and mono-unsaturated fatty acid chains of 6-22 carbons in length. Develops equations for predicting higher heating value from carbon number, electron number, or molecular weight. The second paper includes all data presented in the first.

Bulk Modulus

Varde, K.S. 1984. Bulk Modulus of Vegetable Oil-Diesel Fuel Blends. *Fuel* 63, no. 5 (May): 713-715. dx.doi.org/10.1016/0016-2361(84)90172-8.

Reports bulk modulus for blends of refined soybean oil and petrodiesel blends.

Lubricity

Fernando, Sandun, Milford Hanna and Sushil Adhikari. 2007. Lubricity characteristics of selected vegetable oils, animal fats, and their derivatives. *Applied Engineering in Agriculture* 23, no. 1: 5-11. asae.frymulti.com/abstract.asp?aid=22324.

Vegetable oil and animal fats have better lubricity than petrodiesel, and not just because they are thicker.

Cetane

Freedman, B., M. O. Bagby, T. J. Callahan, and T. W. Ryan III. 1990. Cetane numbers of fatty esters, fatty alcohols and triglycerides determined in a constant volume combustion bomb. *SAE Technical Paper Series:* 900343.

Charts how longer carbon chains have a shorter ignition delay and more highly unsaturated oils and other fatty material have longer ignition delay. Very good work.

Flash, Flame, and Autoignition Points

Wijayasinghe, M., and T. Makey. 1997. Cooking Oil: A Home Fire Hazard in Alberta, Canada. *Fire Technology* 33, no. 2 (May 1): 140-166. dx.doi.org/10.1023/A:1015395001403.

Presents smoke point, flash point, and auto ignition temperatures for seven common cooking oils.

Boiling Points

Goodrum, J. W., and D. P. Geller. 2002. Rapid thermogravimetric measurements of boiling points and vapor pressure of saturated medium- and long-chain triglycerides. *Bioresource Technology* 84, no. 1: 75-80. dx.doi.org/10.1016/S0960-8524(02)00006-8.

Gives boiling points for saturated fatty acids composed of 12, 14, 16, and 18 carbon fatty acids.

Specific Heat

Kasprzycka-Guttman, T. and D. Odzeniak. 1991. Specific heats of some oils and a fat. *Thermochimica Acta* 191, no 1: 41-45. dx.doi.org.proxy.uchicago.edu/10.1016/0040-6031(91)87235-O.

Gives specific heat capacity information for olive, sunflower, soybean, rape, linen, and castor oil and lard.

Morad, N.A., A. A. Mustafa Kamal, F. Panau, and T. W. Yew. 2000. Liquid specific heat capacity estimation for fatty acids, triacylglycerols, and vegetable oils based on their fatty acid composition. *Journal of the American Oil Chemists' Society* 77, no. 9: 1001-1005. dx.doi.org/10.1007/s11746-000-0158-6.

Gives equations to estimate heat capacity based upon fatty acid profile.

Santos, J. C. O., I. M. G. Santos, M. M. Conceicao, S. L. Porto, M. F. S. Trindade, A. G. Souza, S. Prasad, V. J. Fernandes Jr. and A. S. Araujo. 2004. Thermoanalytical, kinetic and rheological parameters of commercial edible vegetable oils. *Journal of Thermal Analysis and Calorimetry* 75, no. 2: 119-128. dx.doi.org/10.1023/B:JTAN.0000027128.62480.db.

Examined thermal decomposition of six vegetable oils. Provides insights into the breakdown of vegetable oils. Good heat capacity data.

Solubility

King, J.W. 1995. Determination of the Solubility Parameter of Soybean Oil by Inverse Gas-Chromatography. *Lebensmittel Wissenschaft & Technologie* 28, no. 2: 190-195. dx.doi.org/10.1016/S0023-6438(95)91398-X

Reports the Hildebrand solubility parameter for soybean oil at various temperatures from 59° to 123°C; through this range the parameter falls from 7.9 to 6.9. $\frac{cal^{1/z}}{cm^{2/z}}$

Hu, Jianbo, Zexue Du, Zhong Tang, and Enze Min. 2004. Study on the solvent power of a new green solvent: Biodiesel. *Industrial and Engineering Chemistry Research* 43, no. 24: 7928-7931. dx.doi.org/10.1021/ie0493816.

Reports Kauri-Butanol values for methyl and ethyl esters made from sunflower, corn, canola, and soybean oil. The values cluster around 80.

Burke, J. *Solubility Parameters: Theory and Application.* palimpsest.stanford.edu/byauth/burke/solpar/solpar3.html.

A very good introduction to solvency.

Cole-Parmer: Chemical Resistance Database. coleparmer.com/techinfo/chemcomp.asp.

Good place to start for information about material compatibility.

National Biodiesel Board. N.d. *Materials Compatibility.* biodiesel.org/pdf_files/fuelfactsheets/Materials_Compatibility.pdf. *Guide to material compatibility of biodiesel.*

Proposed Fuel Standards

Remmele, E. 2007. Pre-Standard DIN V 51605 for Rapeseed Oil Fuel. *15th European Biomass Conference & Exhibition.* *This is specification is very close to becoming an official standard of DIN, the main German standards body.*

Water and Oil

Gaonkar, A. G., and R. P. Borwankar. 1991. Adsorption behavior of monoglycerides at the vegetable oil/water interface. *Journal of Colloid and Interface Science* 146, no. 2: 525-532. dx.doi.org/10.1016/0021-9797(91)90216-U. *Small amounts of monoglyceride (<.05%), can dramatically reduce the interfacial tension between oil and water, allowing stable oil/water mixtures to form much more easily. More monoglycerides further reduce interfacial tension until a floor is reached around 2.5% monoglycerides.*

Xiaohu L., J. Li, and C. Sun. 2006. Properties of Transgenic Rapeseed Oil based Dielectric Liquid. *Proceedings of the IEEE SoutheastCon, 2006.* dx.doi.org/10.1109/second.2006.1629328. *The only reference I've seen to the solubility of water in pure vegetable oil: .8 ppm at 20°C, 2.7 ppm at 90°C.*

Heusch, R. 2002. Emulsions. In *Ullman's Encyclopedia of Industrial Chemistry.* John Wiley & Sons. dx.doi.org/10.1002/14356007.a09_297. *Very thorough introduction to emulsions, a critical topic for using waste vegetable as fuel.*

Straight Vegetable Oil Durability Tests

Hawkins, C. S., J. Fuls, and F. J. C. Hugo. 1983. Engine durability tests with sunflower oil in an indirect injection diesel engine. *SAE Technical Paper Series:* 831357. *Describes modifications made to two Deutz tractors with F3l912W engines that allowed for successful long term (1,800 hour) use of sunflower oil as fuel. The modifications were wrapping a small copper coil around the exhaust pipe to heat the vegetable oil and replacing the diaphragm-type lift pump with a piston-style pump. As a result of the testing, Deutz extended full factory warranties to cover their indirect injected engines fueled with degummed sunflower oil.*

Coelho, S.T., O.C. Silva, S.M.S.G. Velazquez, and M.B.C.A. Monteiro. 2005. "In natura" palm oil as fuel in a conventional diesel engine — implantation and tests at the Vila Soledade community, Para, Brazil. *2005 International Conference on Future Power Systems.* ieeexplore.ieee.org/iel5/10666/33649/01600484.pdf?arnumber=1600484

A generator powered by a MWM TD-229 six cylinder, direct injected engine has run for over 3000 hours on vegetable oil, supplying power to a rural Brazilian village. The engine was fitted with a conversion kit that claimed to heat the oil to 85°C before injection. Every day the engine was run on filtered, crude palm oil for five hours and diesel for one hour to flush the engine. The oil was produced locally.

de Almeida, Silvio C. A. De, Carlos Rodrigues Belchior, Marcos V. G. Nascimento, Leonardo dos S. R. Vieira, and Guilherme Fleury. 2002a. Performance of a Diesel Generator Fueled with Palm Oil. *Fuel* 81, no. 16: 2097-2102. dx.doi.org/10.1016/S0016-2361(02)00155-2.

A MWM 229, six cylinder, direct-injected engine was fed with petrodiesel, 50°C refined palm oil, and 100°C, and run for 100 hours on each fuel. When fueled with 50°C oil, heavy carbon deposits formed on the injector nozzles, but no significant coking was found after 100 hours being fueled by 100°C oil. The researchers concluded that appropriately heated fuel could be used long term in this engine.

Ammerer, A. J. Rathbauer, and M. Worgetter. 2004. Rapeseed Oil as Fuel for Farm Tractors. *IEA Bioenergy Task 39 — Liquid Biofuels.* task39.org/LinkClick.aspx?fileticket=eD329wLXY1w%3d

Long description of the state of the vegetable oil fuel in Germany and Austria, covering production of rapeseed, technical standards, infrastructure, and preliminary results of a state-sponsored "100 tractor" durability test, started in 2000. So far there has been a 5 percent failure rate clearly attributable to vegetable oil fuel, but the report does not specify what set-ups had problems. Worth reading.

Bari, S. 2004a. Investigation into the deteriorated performance of diesel engine after prolonged use of vegetable oil. *Proceedings of the Fall Technical Conference of the ASME International Combustion Engine Division, Oct 24-27 2004*: 447-455.

———. 2002b. Performance deterioration and durability issues while running a diesel engine with crude palm oil. *Proceedings of the Institution of Mechanical Engineers, Part D: Journal of Automobile Engineering* 216, no. 9: 785-792. dx.doi.org/10.1243/09544070260340871.

These two papers describe deteriorated engine performance of a Yanmar L60 AE single cylinder, direct-injection engine after 500 hours of operation being fed by crude palm oil heated to 60°C. carbon deposits were found in the combustion chamber, particularly on the fuel injector, and valve seats and exhaust valve stem. There were traces of wear of the piston rings, upper ring landing, and cylinder lining, as well as wear on the plunger and delivery valve of the injection pump. It should be noted that crude palm oil is one of the thickest vegetable oils and is semi-solid at room temperature.

Pryor, R. W., M. A. Hanna, J. L. Schinstock, and L. L. Bashford. 1983. Soybean Oil Fuel in a Small Diesel Engine. *Transactions of the ASAE* 26, no. 2: 333-337.

Soybean methyl esters, and 57°C crude soybean oil and crude-degummed soybean oil were all acceptable short-term substitutes for a Ford 2600, three cylinder, direct injected tractor. However, the 33-hour durability test for crude soybean oil was cut short by the rupture of a high pressure injection line caused by a seized injector. Besides the seized injector, there was severe coking on all the injector nozzles and the nozzles would no longer spray, but only drip, there were heavy carbon deposits on the piston face and exhaust valves, and pitting on the exhaust valve face. It was concluded that 57°C crude soybean oil cannot be a long-term petrodiesel substitute for this engine.

Karaosmanoglu, F., G. Kurt, and T. Ozaktas. 2000. Direct use of sunflower oil as a compression-ignition engine fuel. *Energy Sources* 22, no. 7: 659-672. dx.doi.org/10.1080/00908310050045618.

———. 2000. Long term CI engine test of sunflower oil. *Renewable Energy* 19, no. 1-2: 219-221.
dx.doi.org/10.1016/S0960-1481(99)00034-8.

Performance, emissions, and medium term (50 hours) durability were evaluated for a Pancar E-108 single cylinder, direct injection engine fueled with unheated, refined sunflower oil. The engine produced acceptable levels of power while using slightly more of the alternative fuel. While fueled with sunflower oil, the engine produced more carbon monoxide, NOx, and hydrocarbons, but less smoke. After the 50-hour durability test showed a small increase in lubricating oil viscosity. The injector nozzle was free from deposits. The second paper merely repeats the findings of the durability test, and that in an abbreviated form.

Clevenger, M. D., M. O. Bagby, C. E. Goering, A. W. Schwab, and L. D. Savage. 1988. Developing an accelerated test of coking tendencies of alternative fuels. *Transactions of the ASAE* 31, no. 4: 1054-1058.

Describes a test to detect engine coking within 5 hours.

Straight Vegetable Oil Performance Tests

Pugazhvadivu, M., and K. Jeyachandran. 2004. Effect of fuel injection pressure and preheating on the performance and emissions of a vegetable oil fuelled diesel engine. *Proceedings of the Fall Technical Conference of the ASME International Combustion Engine Division, Oct 24-27 2004*: 479-486.

———. 2005. Investigations on the performance and exhaust emissions of a diesel engine using preheated waste frying oil as fuel. *Renewable Energy* 30, no. 14: 2189-2202. dx.doi.org/10.1016/j.renene.2005.02.001.

A Kirloskar single cylinder, direct injection engine was fed with waste sunflower oil heated to 30°, 70°, and 135°C. The hotter the oil, the better the engine performed, and the lower the CO and smoke emissions. The difference in performance and emissions between the 70° and 135°C oil was much smaller than the difference between the 30° and 70° oil. Increasing pressure improved performance and reduced smoke, and slightly increased NO_X.

Hebbal, O.D., K. Vijayakumar Reddy, and K. Rajagopal. 2006. Performance characteristics of a diesel engine with deccan hemp oil. *Fuel* 85, no. 14-15 (October): 2187-2194.
dx.doi.org/10.1016/j.fuel.2006.03.011.

Compared the performance and emissions of a Kirloskar AV1 single cylinder, direct injection engine when fueled with petrodiesel, deccan hemp oil, and petrodiesel/vegetable oil blends containing 25 percent, 50 percent, 75 percent deccan hemp oil. The straight oil, and the 50 percent and 75 percent oil blends were heated to temperatures that reduced their viscosity to that of petrodiesel at 30°C. The oil and blends produced similar power per unit energy, greater particulate and carbon monoxide emissions, and higher exhaust temperature emissions. The injector nozzles coked more quickly when running the alternative fuel or blends, but this did not prevent the researcher from concluding that deccan oil could be used as a substitute diesel fuel if it was blended with diesel below 25 percent or if preheated at high concentrations.

Agarwal, D., and A. K. Agarwal. 2007. Performance and emissions characteristics of jatropha oil (preheated and blends) in a direct injection compression ignition engine. *Applied Thermal Engineering* 27, no. 13 (September): 2314-2323.
dx.doi.org/10.1016/j.applthermaleng.2007.01.009.
A Kirloskar single cylinder, direct injected was fed with 80-90°C jatropha oil and various unheated blends of jatropha oil and petrodiesel. The heated oil and unheated blends with less than 30 percent jatropha oil produced acceptable performance and emissions.

Engler, C.R., L.A. Johnson, W.A. Lepori, and C.M. Yarbrough. 1983. Effects of processing and chemical characteristics of plant oils on performance of an indirect-injection diesel engine. *Journal of the American Oil Chemists' Society* 60, no. 8: 1592-1596. dx.doi.org/10.1007/BF02666591.
A Yanmar TS50c, single cylinder, indirect injection engine performed better when fueled with refined sunflower and cottonseed oils than crude oils, but still produced carbon deposits and lubricating oil contamination.

Bari, S., T. H. Lim, and C. W. Yu. 2002. Effects of preheating of crude palm oil (CPO) on injection system, performance and emission of a diesel engine. *Renewable Energy* 27, no. 3: 339-351.
dx.doi.org/10.1016/S0960-1481(02)00010-1.
Compared performance and emissions of a Yanmar L60 AE single cylinder, direct injection engine when fueled with petrodiesel and crude palm oil heated to 60°C. When fueled with the oil, the engine had a 6 percent higher peak pressure and a shorter ignition delay. Both carbon monoxide and NO_x *emissions were higher for the alternative fuel.*

Bari, S., C. W. Yu, and T. H. Lim. 2002. Filter clogging and power loss issues while running a diesel engine with waste cooking oil. *Proceedings of the Institution of Mechanical Engineers, Part D: Journal of Automobile Engineering* 216, no. 12: 993-1001.
dx.doi.org/10.1243/095440702762508245.
In order to avoid filter clogging, it was found to be necessary to heat waste palm oil to 55°C. The researchers found that when oil at this temperature was fed to Yanmar L60 AE single cylinder, direct injection engine, the engine had a shorter ignition, and higher carbon monoxide and NO_x *emissions then petrodiesel*

Yu, C. W., S. Bari, and A. Ameen. 2002. A comparison of combustion characteristics of waste cooking oil with diesel as fuel in a direct injection diesel engine. *Proceedings of the Institution of Mechanical Engineers, Part D: Journal of Automobile Engineering* 216, no. 3: 237-243. dx.doi.org/10.1243/0954407021529066.

A Yanmar 160AE-D, single cylinder, direct injection engine was fed with 70°C waste palm oil. Performance was acceptable, though carbon monoxide, NO_x, and sulfur oxide emissions were higher burning the vegetable oil versus vegetable oil. The ignition delay of the heated palm oil was shorter than petrodiesel.

Nwafor, O. M. I. 1999. Effect of varying fuel inlet temperature on the performance of vegetable oil in a diesel engine under part-load conditions. *International Journal of Ambient Energy* 20, no. 4: 205-210.

———. 2001. Emission characteristics of neat rapeseed oil fuel in diesel engine. *International Journal of Ambient Energy* 22, no. 3 (July): 146-154.

———. 2003. The effect of elevated fuel inlet temperature on performance of diesel engine running on neat vegetable oil at constant speed conditions. *Renewable Energy* 28, no. 2 (February): 171-181. sciencedirect.com/science/article/B6V4S-46PYM5P-2/2/c259577fa5ad492be94947e155769965.

These papers contain very thorough performance analysis of a Petter ACI, single cylinder, indirect injection engine fueled with petrodiesel, unheated rapeseed oil, and 70°C rapeseed oil. Includes pressure crankangle diagrams, and net heat release diagrams which are very interesting. Brake thermal efficiency was found to be higher for the alternative fuels and hydrocarbons much lower. In general, the 70°C oil produced results that were more similar to petrodiesel than the unheated oil.

———. 2004. Emission characteristics of diesel engine running on vegetable oil with elevated fuel inlet temperature. *Biomass and Bioenergy* 27, no. 5: 507-511. dx.doi.org/10.1016/j.biombioe.2004.02.004.

Emissions were compared for a Petter ACI, single cylinder, indirect injection engine burning petrodiesel, unheated rapeseed oil, and 70°C rapeseed oil. Heated rapeseed oil produced more carbon monoxide than petrodiesel during high load conditions, and unheated vegetable oil produced more than either petrodiesel or heated oil under all load conditions. Both heated and unheated vegetable oil produced lower hydrocarbon emissions. Heated vegetable oil had a shorter ignition delay than unheated oil, closer to petrodiesel.

Senthil Kumar, M., A. Kerihuel, J. Bellettre, and M. Tazerout. 2005. Experimental investigations on the use of preheated animal fat as fuel in a compression ignition engine. *Renewable Energy* 30, no. 9 (July): 1443-1456. dx.doi.org/10.1016/j.renene.2004.11.003.

A Lister-Petter TS-1 Series, single cylinder, direct injected engine performed better when fueled with 70°C, animal fat than with 30°C, to the point that the researchers were. Interesting discussion on ignition delays.

Labeckas, G., and S. Slavinskas. 2005. Performance and exhaust emissions of direct-injection diesel engine operating on rapeseed oil and its blends with diesel fuel. *Transport* 20, no. 5: 186-194. transport.vgtu.lt/en/?page=3&pub=1399.

———. 2006. Performance of direct-injection off-road diesel engine on rapeseed oil. *Renewable* Energy 31, no. 6: 849-863. dx.doi.org/10.1016/j.renene.2005.05.009.

These two papers evaluate the performance and emissions of a Minsk Motor Plant (MMZ) D-243, four cylinder, direct injection tractor engine fueled with 60°C, cold pressed rapeseed oil and blends of petrodiesel and vegetable oil. In short term testing, the alternative fuels did well.

Altin, R., S. Cetinkaya, and H. S. Yucesu. 2001. The potential of using vegetable oil fuels as fuel for diesel engines. *Energy Conversion and Management* 42, no. 5 (March): 529-538. dx.doi.org/10.1016/S0196-8904(00)00080-7.

Compared performance and emissions of raw sunflower, cottonseed, and soybean oil; processed corn poppy, and rapeseed oil; methyl esters of sunflower, cottonseed, and soybean oil, and petrodiesel, in a Superstar 7710 single cylinder, direct injected engine. The oils were heated to 80°C. All the alternate fuels performed acceptably, without any real differences between fuels in torque or fuel consumption. Similarly, the alternate fuels produced less NO_x and more particulate matter emissions than petrodiesel, without clear differences emerging between the alternate fuels. The straight vegetable oils did produce more carbon monoxide than the methyl esters or petrodiesel.

Blend Durability Test

Adams, C., J.F. Peters, M.C. Rand, B.J. Schroer, and M.C. Ziemke. 1983. Investigation of soybean oil as a diesel fuel extender: Endurance tests. *Journal of the American Oil Chemists' Society* 60, no. 8: 1574-1579. dx.doi.org/10.1007/BF02666588.

A 50/50 blend of refined soybean oil and petrodiesel caused unacceptable increases in the viscosity of engine lubricating oil of a John Deere 6404TR, six cylinder, direct injection during a 200-hour durability test. A blend of 33/66 soybean oil and petrodiesel did not cause problems with engine oil viscosity.

Ziejewski, M., and K. R. Kaufman. 1983. Laboratory endurance test of a sunflower oil blend in a diesel engine. *Journal of the American Oil Chemists' Society* 60, no. 8: 1567-1573. dx.doi.org/10.1007/BF02666587.

An Allis-Chalmers, four cylinder, direct-injected engine failed a long-term (200 hr) durability test when fueled with a 25/75 blend of refined sunflower oil and diesel. Compared to the baseline fuel, there were heavy deposits on the injector nozzles, exhaust valve stems, cylinder sleeves, and piston ring landing.

Ziejewski, M., and K. R. Kaufman. 1983. Laboratory endurance test of a sunflower oil blend in a diesel engine. *Journal of the American Oil Chemists' Society* 60, no. 8: 1567-1573. dx.doi.org/10.1007/BF02666587.

Ziejewski, M., K. R. Kaufmann, A. W. Schwab, and E. H. Pryde. 1983. Diesel engine evaluation of a nonionic sunflower oil-aqueous ethanol microemulsion.*Journal of the American Oil Chemists' Society* 61, no. 10: 1620-1626. dx.doi.org/10.1007/BF02541646.

Blend Peformance Tests

Rakopoulos, C.D., K.A. Antonopoulos, D.C. Rakopoulos, D.T. Hountalas and E.G. Giakoumis. 2006. Comparative performance and emissions study of a direct injection diesel engine using blends of diesel fuel with vegetable oils or bio-diesels of various origins. *Energy Conversion and Management* 47, no 18-19:3372-3287. dx.doi.org/10.1016/j.enconman.2006.01.006.

A very sensitive discussion of the emission and performance of a Ricardo/Cussons 'Hydra' single-cylinder engine operating in direct-injected mode fueled with petrodiesel; methyl esters of cottonseed, soybean, sunflower, rapeseed, and palm oil; and refined and degummed cottonseed, sunflower, corn, and olive kernel oil. The biofuels were blended at 10 percent and 20 percent with petrodiesel. The paper is very good, not so much for its results, which are not surprising, but for the excellent theoretical discussion of how the biofuels could affect emissions.

Ozaktas, T., K. B. Cigizoglu, and F. Karaosmanoglu. 1997. Alternative diesel fuel study on four different types of vegetable oils of Turkish origin. *Energy Sources* 19, no. 2: 173-181. dx.doi.org/10.1080/00908319708908842.

A 20/80 blend of refined sunflower, olive, soybean, and corn oil and petrodiesel were used to fuel a Daimler Benz OM, six cylinder, indirect engine. In the short term tests (30 minutes), the engine performed comparably to when it was fueled with straight petrodiesel, and produced less smoke.

Cigizoglu, K. B., T. Ozaktas, and F. Karaosmanoglu. 1997. Used sunflower oil as an alternative fuel for diesel engines. *Energy Sources* 19, no. 6: 559-566. dx.doi.org/10.1080/00908319708908872.

A 20/80 blend of used sunflower oil and petrodiesel were used to fuel a Daimler Benz OM, six cylinder, indirect engine. In the short term tests (30 minutes), the engine performed comparably to when it was fueled with straight petrodiesel, and produced less smoke.

Ozaktas, T. 2000. Compression ignition engine fuel properties of a used sunflower oil-diesel blend. *Energy Sources* 22:377-382. dx.doi.org/10.1080/00908310050013974.

Primarily interesting for providing fuel properties for used *sunflower oil, however, the value for viscosity seems way off, so the value of the rest of data is questionable.*

Huzayyin, A. S., A. H. Bawady, and M. A. Rady. 2004. Experimental evaluation of diesel engine performance and emissions using blends of jojoba oil and diesel fuel. *Energy Conversion and Management* 45, no 13-14:2093-2112. dx.doi.org/10.1016/j.enconman.2003.10.017.

In a short-term test, A Deutz F1l511 single cylinder, direct injection engine performed acceptable and with lower NO_x *and soot emission when fueled with up to 60 percent filtered, crude jojoba oil.*

Pramanik, K. 2003. Properties and use of jatropha curcas oil and diesel fuel blends in compression ignition engines. *Renewable Energy* 28, no 2:239:248. dx.doi.org/10.1016/S0960-1481(02)00027-7.

A Kirloskar single cylinder, direct injection engine was fueled with a variety of blends of jatropha curcas oil and petrodiesel. Short term tests showed that the engine performed acceptable when fueled with blends containing up to 50 percent vegetable oil.

Bartholomew, D. 1981. Viewpoint: vegetable oil fuel. *Journal of the American Oil Chemists' Society* 58, no. 4: 286-288 dx.doi.org/10.1007/BF02541575.

A view from a Merrill Lynch oil trader on interest in vegetable oil as fuel. References Caterpillar extending warranty to vegetable oil blends in Brazil.

Emissions

Lance, D. L., and J. D. Andersson. 2004. Emissions performance of pure vegetable oil in two European light duty vehicles. *SAE Technical Paper Series:* 2004-01-1881.

Analyzes the emissions of two unspecified British vehicles, one with a 1.9 liter direct injection engine, and the other with a 1.5 indirect injection engine. Fitted with an unspecified commercial conversion kit that the authors claim maintains the fuel fed to the engine at 80°C, but do not seem to have verified. The vehicles were fueled with ultra low sulfur diesel, a 5 percent blend of rapeseed methyl esters and ultra low sulfur diesel, and refined, food grade canola oil. When the direct injected vehicle was fueled with the canola, the authors reported much higher carbon monoxide, particulate emissions, and slightly lower NO_X. For the indirect injected vehicle, carbon monoxide and hydrocarbon emissions were higher, and particulate and NO_X emissions were lower. Closer examination of the exhaust showed that burning the canola oil produced much higher levels of polycyclical aromatic hydrocarbons which are a class of compounds that are considered mutanegenic and carcinogenic. This paper has been widely criticized by proponents of vegetable oil as diesel fuel for not specifying the vehicles, and not clearly verifying the injection temperatures of the vegetable oil.

Agarwal, D., S. Sinha, and A. Agarwal. 2006. Experimental investigation of control of NO_X emissions in biodiesel-fueled compression ignition engine. *Renewable Energy* 13, no. 14: 2356-2369. dx.doi.org.proxy.lib.utk.edu:90/10.1016/j.renene.2005.12.003.

Exhaust Gas Recirculation was shown reduce NO_X emissions of an engine burning various blends of petrodiesel and biodiesel without a significant penalty in particulate matter emissions or fuel consumption.

Monyem, A., J. H. Van Gerpen, and M. Canakci. 2001. The effect of timing and oxidation on emissions from biodiesel-fueled engines. *Transactions of the American Society of Agricultural Engineers 44, no. 1: 35-42. asae.frymulti.com/abstract.asp?aid=2301.*

Researchers showed that retarding the timing of a John Deere 4276T four cylinder, direct injection engine would result in lower NO_X emissions.

Spray, Ignition, and Precombustion

Callahan, T. J., T. W. Ryan III, L. G. Dodge, and J. A. Schwalb. 1988. Effect of Fuel Properties on Diesel Spray Characteristics. *SAE Technical Paper Series:* 870533.

Using special equipment built to reproduce the pressures of a diesel engine cylinder, the spray pattern of four petroleum-based fuels with different specific gravities and viscosities were compared, and it was found that more viscous fuel produced a spray that was narrower and held together longer before breaking apart.

Yoshimoto, Y. 2003. Performance and Emissions of Diesel Fuels Containing Rapeseed Oil and the Characteristics of Evaporation and Combustion of Single Droplets. *SAE Technical Paper Series:* 2003-01-3201.

Examines how individual droplets of rapeseed/diesel blends evaporate and combust.

Jozwiak, D., and A. Szlek. 2007a. Ignition Characteristics of Vegetable Fuel Oils in Fuel Spray. *Journal of the Energy Institute* 80, no. 1: 35-39. dx.doi.org/10.1179/174602207X174360.

Examines the role of cylinder temperature, droplet size, and percentages of rapeseed oil/petrodiesel in affecting the ignition delay of individual droplets of rapseed oil and petrodiesel blends.

Knothe, G., M. O. Bagby, T. W. Ryan III, and T. J. Callahan. 1991. Degradation of Unsaturated Triglycerides Injected into a Pressurized Reactor. *Journal of the American Oil Chemists' Society* 68, no. 4: 259-267. dx.doi.org/10.1007/BF02657621.

Knothe, G., M. O. Bagby, T. W. Ryan III, T. J. Callahan, and H. G. Wheeler. 1992. Vegetable Oils as Alternative Diesel Fuels: Degradation of Pure Triglycerides during the Precombustion Phase in a Reactor Simulating a Diesel Engine. *SAE Technical Paper Series:* 920194.

Knothe, G., M. O. Bagby, T. W. Ryan III, H. G. Wheeler, and T. J. Callahan. 1992. Semi-volatile and Volatile Compounds Formed by Degradation of Triglycerides in a Pressurized Reactor. *Journal of the American Oil Chemists' Society* 69, no. 4 (April 16): 341-346. dx.doi.org/10.1007/BF02636064.

Ryan, T. W. III, and M. O. Bagby. 1993. Identification of Chemical Changes Occurring During the Transient Injection of Selected Vegetable Oils. *SAE Technical Paper Series:* 930933.

Knothe, G., M. O. Bagby, and T. W. Ryan III. 1998. Precombustion of Fatty Acids and Esters of Biodiesel. A Possible Explanation for

Differing Cetane numbers. *Journal of the American Oil Chemists' Society* 75, no. 8: 1007-1013. dx.doi.org/10.1007/s11746-998-0279-1.
These five papers examine the chemical reactions that occur before ignition, injecting vegetable oil fuel into a cylinder with comparable pressure and temperature to that of combustion cylinder, and that has been evacuated of oxygen.

Engine Modifications

Prasad, C. M. V., M. V. S. Krishna, C. P. Reddy, and K. R. Mohan. 2000. Performance Evaluation of Non-edible Vegetable Oils as Substitute Fuels in Low Heat Rejection Diesel Engines. *Proceedings of the Institution of Mechanical Engineers, Part D: Journal of Automobile Engineering* 214, no. 2: 181-187.
dx.doi.org/10.1243/0954407001527330.
Heating the oil, increasing injection pressure, and using an insulated piston and cylinder liner all individually improved the performance of a Kirloskar single cylinder, direct injection engine burning Pongamia and Jatropha curcas oil.

Nwafor, O. M. I., G. Rice, and A. I. Ogbonna. 2000. Effect of Advanced Injection Timing on the Performance of Rapeseed Oil in Diesel Engines. *Renewable Energy* 21, no. 3: 433-444.
dx.doi.org/10.1016/S0960-1481(00)00037-9.
Advancing injection timing increased exhaust temperatures, decreased carbon monoxide. Strangely, the ignition delay for vegetable oil decreased with advanced injection. NO_X *was not measured.*

Prasad, G. A. P. and P. R. Mohan. 2003. Effect of Supercharging on the Performance of a DI Diesel Engine with Cottonseed Oil. *Energy Conversion and Management*, 44:937–944.
Supercharging a naturally aspirated Kirloskar AVI single cylinder, direct injected engine improved the performance when fueled with unheated cottonseed oil, decreasing fuel consumption and reducing smoke emissions. Increasing injection pressure did not have much effect.

Raja, A.Samuel, G. Lakshmi Narayana Rao, N. Nallusamy, and M. Selva Ganesh Kumar. 2004. Effect of Combustion Chamber Design on the Performance and Emission Characteristics of a Direct Injection (Dl) Diesel Engine Fueled with Refined Rice Bran Oil Blends. *Proceedings of the Fall Technical Conference of the ASME International Combustion Engine Division, Oct 24-27 2004*: 427-435.

The performance and emissions of various refined rice oil/diesel blends were evaluated in single cylinder, direct injection engine that was fitted with three different piston designs, one with spherical combustion chamber, one with a re-entrant combustion chamber, and the last with a torroidal combustion chamber. The results were not conclusive, and the data is presented in a way that makes it difficult to directly compare different piston designs.

Bari, S., C. W. Yu, and T. H. Lim. 2004. Effect of Fuel Injection Timing with Waste Cooking Oil as a Fuel in a Direct Injection Diesel Engine. *Proceedings of the Institution of Mechanical Engineers, Part D: Journal of Automobile Engineering* 218, no. 1: 93-104. dx.doi.org/10.1243/095440704322829209.

Advancing the timing 4° in a Yanmar L60 AE single cylinder, direct injection engine increased efficiency by 1.6 percent but also increased NO_X emissions 76.6 percent. When the engine was fueled with petrodiesel, the advance in timing produced a smaller increase in efficiency and a larger increase in NO_X.

Bannikov, M.G., S.I. Tyrlovoy, I.P. Vasilev, and J.A. Chattha. 2006. Investigation of the Characteristics of the Fuel Injection Pump of a Diesel Engine Fueled with Viscous Vegetable Oil-Diesel Oil Blends. *Proceedings of the Institution of Mechanical Engineers, Part D: Journal of Automobile Engineering* 220, no. 6: 787-792. dx.doi.org/10.1243/09544070JAUTO88.

Suggests alterations that may be made to a Bosch VE-type injection pump in order compensate for the different physical properties of vegetable oil and produce a torque curve on the alternative fuel that matches that of the base fuel. The alterations suggested are intensive, and not immediately practicable.

Modeling

These articles will be of interest to those wishing to model the combustion of vegetable oil in a diesel engine.

Radu, R., E. Rakosi, C. Iulian-Agape, and R. Gaiginschi. 2001. Application of a Combustion Model to a Diesel Engine Fueled with Vegetable Oils. *JSME International Journal, Series B: Fluids and Thermal Engineering* 44, no. 4: 634-640. dx.doi.org/10.1299/jsmeb.44.634.

Rakopoulos, C.D., K.A. Antonopoulos, and D.C. Rakopoulos. 2006. Multi-zone Modeling of Diesel Engine Fuel Spray Development with Vegetable Oil, Bio-Diesel or Diesel Fuels. *Energy Conversion and*

Management 47, no. 11-12 (July): 1550-1573.
dx.doi.org/10.1016/j.enconman.2005.08.005.

Raubold, W. 1995. Thermodynamic analysis of the engine internal process to determine the suitability of vegetable oils as alternative fuels for diesel engines. *Proceedings of the ASME Internal Combustion Engine Division Spring Meeting, Apr 23-26 1995.*

Griend, L.V., M.E. Feldman, and C.L. Peterson. 1990. Modeling Combustion of Alternate Fuels in DI Diesel Engine Using KIVA. *Transactions of the ASAE* 33, no. 2: 342-350.

Yuan, W., A.C. Hansen, M. E. Tat, J. H. Van Gerpen, and Z. Tan. 2005. Spray, ignition, and combustion modeling of biodiesel fuels for investigating NO_X emissions. *Transactions of the American Society of Agricultural Engineers* 48, no. 3: 933-939.
asae.frymulti.com/abstract.asp?aid=18498.

Msipa, C. K. M., C. E. Goering, and T. D. Karcher. 1983. Vegetable oil atomization in a DI diesel engine. *Transactions of the ASAE* 26, no. 6: 1669-1672.

Small Scale Production

Helgeson, D. L., and L. W. Schaffner. 1983. Economics of on-farm processing of sunflower oil. *Journal of the American Oil Chemists' Society* 60, no. 8: 1561-1566. dx.doi.org/10.1007/BF02666586.

Backer, L. F., L. Jacobsen, and J. C. Olson. 1983. Farm-scale recovery and filtration characteristics of sunflower oil for us in diesel engines. *Journal of the American Oil Chemists' Society* 60, no. 8: 1558-1560.
dx.doi.org/10.1007/BF02666585

Strayer, R.C., J.A. Blake, and W.K. Craig. 1983. Canola and high erucic rapeseed oil as substitutes for diesel fuel: Preliminary tests. *Journal of the American Oil Chemists' Society* 60, no. 8: 1587-1592.
dx.doi.org/10.1007/BF02666590.
Generally positive research on small-scale oil production and short term testing on a Petter two-cylinder direct injection engine and a John Deere six-cylinder, direct injection engine fueled with canola and high erucic rapeseed oils.

Development of Low Molecular Weight Oil Crops

These papers detail research done to produce oilseed crops that produce lower weight oils, thinner viscosity oils that can be used directly in diesel engines.

Geller, D. P., and J. W. Goodrum. 2000. Rheology of vegetable oil analogs and triglycerides. *Journal of the American Oil Chemists' Society* 77, no. 2: 111-114.
dx.doi.org/10.1007/s11746-000-0018-4.

Geller, D. P., J. W. Goodrum, and C. C. Campbell. 1999. Rapid screening of biologically modified vegetable oils for fuel performance. *Transactions of the ASAE* 42, no. 4: 859-862.
asae.frymulti.com/abstract.asp?aid=13264.

Geller, D. P., J. W. Goodrum, and E. A. Siesel. 2003. Atomization of short-chain triglycerides and a low molecular weight vegetable oil analogue in DI diesel engines. *Transactions of the American Society of Agricultural Engineers* 46, no. 4: 955-958.
asae.frymulti.com/abstract.asp?aid=13950.

Goodrum, J. W., and M. A. Eiteman. 1996. Physical properties of low molecular weight triglycerides for the development of bio-diesel fuel models. *Bioresource Technology* 56, no. 1: 55-60.
dx.doi.org/10.1016/0960-8524(95)00167-0.

Goodrum, J. W., D. P. Geller, and S. A. Lee. 1998. Rapid measurement of boiling points and vapor pressure of binary mixtures of short-chain triglycerides by TGA method. *Thermochimica Acta* 311, no. 1: 71.
dx.doi.org/10.1016/S0040-6031(97)00377-8.

Goodrum, J.W. 1997. Rapid measurements of boiling point and vapor pressure of short-chain triglycerides by thermogravimetric analysis. *Journal of the American Oil Chemists' Society* 74, no. 8: 947-950.
dx.doi.org/10.1007/s11746-997-0009-0.

Goodrum, J. W., and M. A. Eiteman. 1996. Physical properties of low molecular weight triglycerides for the development of bio-diesel fuel models. *Bioresource Technology* 56, no. 1: 55-60.
dx.doi.org/10.1016/0960-8524(95)00167-0

Eiteman, M.A., and J.W. Goodrum. 1994. Density and viscosity of low-molecular weight triglycerides and their mixtures. *Journal of the American Oil Chemists' Society* 71, no. 11: 1261-1265.
dx.doi.org/10.1007/BF02540548.

Goodrum, J. W., V.C. Patel, and R. W. McClendon. 1996. Diesel Injector Carbonization by Three Alternative Fuels. *Transactions of the ASAE* 39, no. 3: 817-821.

Elsbett

Mellde, R. W., I. M. Maasing, and T. B. Johansson. 1989. Advanced automobile engines for fuel economy low emissions, and multifuel capability. *Annual Review of Energy* 14:425-44. dx.doi.org/doi:10.1146/annurev.eg.14.110189.002233.

A good engineering description of the Elsbett engine.

Tang, T.S., H.J. Ahmad Hitam, and Y. Basiron. 1995. Emission of Elsbett Engine Using Palm Oil Fuel. *Journal of Oil Palm Research* 7, no. Special Issue: 110-120. palmoilis.mpob.gov.my/publications/jopr1995sp-110.html.

Analyzes the emissions from a Mercedes 190D fitted with an Elsbett, three cylinder, direct injection engine, fueled with either petrodiesel or palm oil. Burning the vegetable oil, hydrocarbons, NO_X, particulate matters, and a few classes of polcyclical aromatic hydrocarbons were lower, although carbon monoxide was higher.

Engine Oil

Acaroglu, M., H. Oguz, and H. Ogut. 2001a. An investigation of the use of rapeseed oil in agricultural tractors as engine oil. *Energy Sources* 23, no. 9: 823-830. dx.doi.org/10.1080/009083101316931898.

Canola oil was used as engine lubricating oil in a Turkish Tumosan 3D-29DT tractor engine for 50 hours. The oil was an acceptable lubricant during this period, but showed signs of chemical degradation, indicating long-term use was not acceptable without proper stabilizing additives.

Oxidation and Stability

Southwest Research Institute. 2005. *Characterization of biodiesel oxidation and oxidation products*. NREL. nrel.gov/vehiclesandfuels/npbf/pdfs/39096.pdf.

Excellent literature review of the problem of oxidation of biodiesel, much of which is directly applicable to vegetable oil.

Infoletter 09/06: Soy bean oil as a fuel. 2006. ASG Analytik. asg-analytik.de/Downloads/Infobriefe/Infobrief_2006_09_E.pdf.

Provides high quality, color photographs of the kind of damage that can be caused to injectors, pistons, and engine oil by using vegetable oil fuel. The conclusion of the report, that this damage was caused by fueling with soybean oil, as opposed to rapeseed oil, is completely unsupported. The report also claims that soybean oil contains 32 percent more linolenic fatty acids than rapeseed oil. This is not true, rapeseed contains more linolenic fatty acids than soybean. Soybean does contain more linoleic fatty acids.

Beatty, J. 2007. Vegetable Oil as Fuel. frybrid.com/forum/attachment.php?attachmentid=1491&d=1167782933

Written by a commercial vegetable oil chemist and SVO enthusiast, this report describes oxidation in depth, and reports experimental data on effects of metals and a powerful anti-oxidant on rates of oxidation, analysis of moisture, peroxide, and free fatty acid value of samples of oil from the SVO community, the relation between free fatty acids and water. A very important read.

Fox, N.J., and G.W. Stachowiak. 2007. Vegetable oil-based lubricants — A review of oxidation. *Tribology International* 40, no. 7 (July): 1035-1046. dx.doi.org/10.1016/j.triboint.2006.10.001.

Good introduction to oxidation generally and the problem of lubricating oil degradation particularly.

List, G.R., T. Wang, and V.K.S. Shukka. 2005. Storage, handling, and transport, or oils and fats. In *Bailey's Industrial Oil and Fat Products*, ed. F. Shahidi, 191-229. John Wiley & Sons. knovel.com/knovel2/Toc.jsp?BookID=1432&VerticalID=0.

Schaich, K.M. 2005. Lipid oxidation: theoretical aspects. In *Bailey's Industrial Oil and Fat Products*, ed. F. Shahidi, 269-355. John Wiley & Sons. knovel.com/knovel2/Toc.jsp?BookID=1432&VerticalID=0.

The first article from Bailey's presents a great deal of information helpful for preventing oxidation of vegetable oils. The second chapter is a very involved chemical description, and was, I must admit, over my head.

Wheeler, D. H.. 1932. Peroxide formation as a measure of autoxidative deterioration. *Journal of the American Oil Chemists' Society* 9, no. 4 (April 8): 89-97. dx.doi.org/10.1007/BF02553782.

The peroxide test for oxidation is described, and peroxide values of cottonseed oil are compared to iodine number, oil color, and Kreis color.

Holman, R. T., and O. C. Elmer. 1947. The Rates of Oxidation of Unsaturated Fatty Acids and Esters. *The Journal of the American Oil Chemists' Society* 24:127. dx.doi.org/10.1007/BF02643258.

Demonstrates how more unsaturated fatty acids are much more oxidatively unstable.

Wexler, H. 1964. Polymerization of drying oils. *Chemical Reviews* 64, no. 6: 591-611. dx.doi.org/10.1021/cr60232a001.

Fairly good review article on the process of oxidative polymerization. Very technical.

Canakci, M., A. Monyem, and J. H. Van Gerpen. 1999. Accelerated oxidation processes in biodiesel. *Transactions of the ASAE* 42, no. 6: 1565-1572. asae.frymulti.com/abstract.asp?aid=13321

Biodiesel made from soybean oil will oxidize rapidly at temperatures typical of diesel engines fuel tanks (60°C).

Crapiste, G., M. I. V. Brevedan, and A. A. Carelli. 1999. Oxidation of sunflower oil during storage. *Journal of the American Oil Chemists' Society* 76, no. 12 (December 23): 1437-1443. dx.doi.org/10.1007/s11746-999-0181-5.

Rate of oxidation is dependent upon temperature, oxygen availability, and surface area/volume ratio.

Going. 1968. Oxidative deterioration of partially processed soybean oil. *Journal of the American Oil Chemists' Society* 45, no. 9: 632-634. dx.doi.org/10.1007/BF02668968.

Processed oils oxidize more rapidly than crude oils. Rate of oxidation is dependent upon surface area/volume ratio.

Miyashita, K., and T. Takagi. 1986. Study on the oxidative rate and prooxidant activity of free fatty acids. *Journal of the American Oil Chemists' Society* 63, no. 10: 1380-1384. dx.doi.org/10.1007/BF02679607.

Free fatty acids oxidize more quickly than their methyl esters.

Frega, N., M. Mozzon, and G. Lercker. 1999. Effects of free fatty acids on oxidative stability of vegetable oil. *Journal of the American Oil Chemists' Society* 76, no. 3: 325-329. dx.doi.org/10.1007/s11746-999-0239-4.

Free fatty acids accelerated oxidative breakdown of filtered oils, but did not seem to affect unfiltered crude oils.

Fernando, S., and M. Hanna. 2002. Oxidation Characteristics of Soybean Oils as Water Pump Lubricants. *Transactions of the American Society of Agricultural Engineers* 45, no. 6: 1715-1719. asae.frymulti.com/abstract.asp?aid=11417.

Contains good pictures of polymer deposits and decent review of methods to inhibit oxidative polymerization.

Monyem, A., and J. H. Van Gerpen. 2001. The effect of biodiesel oxidation on engine performance and emissions. *Biomass and Bioenergy* 20, no. 4: 317-325. dx.doi.org/10.1016/S0961-9534(00)00095-7.

Biodiesel was mildly oxidized, and then used to fuel a John Deere 4276T four cylinder, direct injection engine, the oxidized biodiesel produced lower carbon monoxide and hydrocarbon emissions compared to unoxidized biodiesel. There was no significant difference in power, fuel consumption, NO_X or smoke.

Knothe, G., and R. O. Dunn. 2003. Dependence of Oil Stability Index of Fatty Compounds on Their Structure and Concentration and Presence of Metals. *Journal of the American Oil Chemists' Society* 80, no. 10: 1021-1026. dx.doi.org/10.1007/s11746-003-0814-x.

Argues for the OSI test for indicating oxidative stability of fatty compounds over the use of Iodine Value. Currently, Iodine Value is used as a standard of oxidative stability in European definitions of biodiesel, and the current standard has the effect of excluding American biodiesel from European markets since most American biodiesel is made from soybean oil which has a relatively high Iodine Value. This article is part of the political fight to open European markets to American biodiesel.

Frankel, E.N. 2005. *Lipid Oxidation.* Bridgewater: P.J. Barnes & Associates.

A book on oxidation of fatty material, written at a very high technical level, and probably of limited usefulness to those not familiar with organic chemistry.

Effect of Frying

Danowska-Oziewicz, M., and M. Karpinska-Tymoszczyk. 2005. Quality Changes in Selected Frying Fats During Heating in a Model System. *Journal of Food Lipids* 12:159-168. dx.doi.org/10.1111/j.1745-4522.2005.00014.x.

Rapeseed and soybean oil along with shortening were heated to 180°C in a deep fat fryer for two hours, two times a day, for six days. Every two hours, the oil was checked for acid value, peroxide value, carbonyl value, and TBA value. Useful, partial model for what may have happened to restaurant oil before discard. Study did not include placing food or other sources of moisture or contaminants in the oil.

Rossell, J.B., ed. 2001. *Frying - Improving Quality.* Woodhead Publishing. knovel.com/knovel2/Toc.jsp?BookID=544&VerticalID=0.

Good overview of cooking oil from industry insider. Particularly good discussion of issues affecting oil quality and storage.

Miscellaneous

Nagao, F., M. Ikegami, and A. Tokunaga. 1966. Temperature dependence of carbon deposits in a diesel combustion chamber. *The Japan Society of*

Mechanical Engineers. 9, no. 35:573-579.

Describes the upper and lower temperature bounds for formation of carbon deposits.

Schwab, A. W., M. O. Bagby, and B. Freedman. 1987. Preparation and properties of diesel fuels from vegetable oils. *Fuel* 66, no. 10 (October): 1372-1378. dx.doi.org/10.1016/0016-2361(87)90184-0.

Interesting paper that describes fuels made from vegetable oil through dilution, microemulsions, pyrolysis, and transesterfication.

Sheehan, J., V. Camobreco, M. Duffield, and H. Shapouri. 1998. *Life Cycle Inventory of Biodiesel and Petroleum Diesel for Use in an Urban Bus*. Golden, Colorado: NREL.

nrel.gov/docs/legosti/fy98/24089.pdf.

Calculates the carbon and energy inputs of biodiesel from planting to tailpipe. Easily extended to vegetable oil.

Books about Diesel Engines

Stone, R. 1999. *Introduction to Internal Combustion Engines*. Warrendale: SAE International.

A great introduction and reference to the principles and operation of diesel engines.

GmbH, Robert Bosch. 2006. *Diesel-Engine Management*. Sussex: John Wiley.

Detailed description of the theory and function of components that are in most diesel engines, from the manufacturer of those components.

Contents of *Vegetable Oil Fuels: Proceedings of the International Conference on Plant and Vegetable Oils as Fuels.*

Oilseed production

- Vegetable oils and animal fats for diesel fuels: a systems study
- Comparison of oilseed yields: a preliminary review
- Energy and economic efficiency for off-site processing of selected oilseeds
- Sunflower production, harvesting, drying and storage
- Double-crop sunflowers for agricultural diesel in the eastern cornbelt
- Analysis of vegetable oil production in central Iowa
- Buffalo gourd: potential as a fuel resource on semi-arid lands

Fuel Preparation/Specification for plant and vegetable oils

- Characterization of vegetable oils for use as fuels in diesel engines
- Sunflower oil as a fuel for compression ignition engines
- Production and fuel characteristics of vegetable oil from oilseed crops in the Pacific Northwest
- Vegetable oil fuel standards

Fuel preparation/specifications for modified plant and vegetable oils

- Vegetable oils — a new alternative
- Fatty esters from vegetable oils for use as a diesel fuel
- Sunflower methyl ester as a diesel fuel
- Transesterification of vegetable oils for fuels

Economics of plant and vegetable oils for fuels

- National implications of substituting plant oils for diesel fuel
- Economic implications for the potential development of a vegetable oil fuel industry
- Oilseed and other biomass fuel penetration in the agricultural sector under alternative price assumptions for competing fuels
- Economics of on-farm sunflower processing
- The economics of on-farm production and use of vegetable oils for fuel

Engine tests with modified plant and vegetable oils

- Comparative combustion studies on various plant oil esters and the long term effect of an ethyl ester on a compression ignition engine
- Methylesters of plant oils as diesel fuels, either straight or in blends

Fuel additives/thermal polymerization

- The influence of lubricant contamination by methyl esters of plant oils on oxidation stability and life
- Polymerization of vegetable oils
- Fuel additives for vegetable oil-fueled compression ignition engines
- Injector-fouling propensity of certain vegetable oils and derivatives as fuels for diesel engines

Oilseed presses and extraction

- Evaluation of an on-farm press
- On-farm soybean oil expression
- An automated small scale oil seed processing plant for production of fuel for diesel engines
- Expellor extracted rape and safflower oilseed meals for poultry and sheep

Short term engine performance

- Evaluation of soybean-aqueous ethanol microemulsions for diesel engines
- Properties and performance testing with blends of biomass alcohols, vegetable oils, and diesel fuel
- Alternative fuel blends and diesel engine tests
- Some correlations of diesel engine performance with injection characteristics using vegetable oil as fuel
- Performance durability and low temperature evaluation of sunflower oil as a diesel fuel extender

Long term durability tests

- Performance of winter rape (Brassica Napus) based fuel mixtures in diesel engines
- Swedish tests on rape seed oil as an alternative to diesel fuel
- Engine deposits and pour point studies using canola oil as a diesel fuel
- Laboratory endurance tests of a sunflower oil blend in a diesel engine
- Performance and durability of a turbocharged diesel fueled with cottonseed oil blends
- The 1981 "Flower Power" field testing program
- First results with a Mercedes-Benz DI diesel engines running on monoesters of vegetable oils

Notes

Introduction

1. J. Sheehan et al., *Life Cycle Inventory of Biodiesel and Petroleum Diesel for Use in an Urban Bus* (Golden, Colorado: NREL, 1998), nrel.gov/docs/legosti/fy98/24089.pdf (accessed September 30, 2007).
2. Vegetable oil and biodiesel have very similar energy contents.

Chapter 1

1. Silvio, C. A., De Almeida et al., "Performance of a diesel generator fueled with palm oil," *Fuel* 81, no. 16 (2002), dx.doi.org/10.1016/S0016-2361(02)00155-2;
   Bari, S., "Investigation into the deteriorated performance of diesel engine after prolonged use of vegetable oil," *Proceedings of the ICEF04, Oct 24-27, 2004*, Paper number:ICEF2004-955, 2004;
   Wagner, G.L., and C. L. Peterson, "Performance of Winter Rape (Brassica Napus) Based Fuel Mixtures in Diesel Engines," in *Vegetable Oil Fuels: Proceedings of the International Conference on Plant and Vegetable Oils as Fuels* (St. Joseph, MI: American Society of Agricultural Engineers, 1982), 329-336;
   Engler, C.R. et al., "Effects of processing and chemical characteristics of plant oils on performance of an indirect-injection diesel engine," *Journal of the American Oil Chemists' Society* 60, no. 8 (1983), dx.doi.org/10.1007/BF02666591;

Pryor, R.W. et al., "Soybean Oil Fuel in a Small Diesel Engine," Transactions of the ASAE 26, no. 2 (1983)

2. Bosch, R, GmbH, *Diesel-Engine Management* (Wiley, 2006): 67-68
3. Stone, R., *Introduction to Internal Combustion Engines* (SAE International, 1999): 226
4. Ibid: 217
5. Brady, R. N., *Modern Diesel Technology* (Prentice Hall, 1995): 10

Chapter 2

1. If a fatty acid has one double bond, it will appear between the 9th and 10th carbon atoms, counting from the carbon that has a double bond with oxygen. If there are two double bonds, they will appear between the 9th and 10th and the 12th and 13th carbon atoms. If there are three, they will appear between the 9th and 10th, 12th and 13th, and 15th and 16th carbon atoms.
2. Five companies produce around 95 percent of our cooking oil: in order of size — ADM, Cargill, Bunge, AGP, and Harvest State.
3. Kantonales Laboratorium BS - Berichte. kantonslabor-bs.ch/content.cfm?nav=17&content=23&Command=details&year=2004&kat=all&ID=75.
4. Alcohols are carbon-based compounds that have – OH groups.
5. Bhattacharyya, S., and C. S. Reddy. 1994. Vegetable Oils as Fuels for Internal Combustion Engines: A Review. *Journal of Agricultural Engineering Research* 57, no. 3 (March): 157-166. dx.doi.org/10.1006/jaer.1994.1015.
6. Goering, C. E., A. Schwab, M. Dougherty, M. Pryde, and A. Heakin. 1982. Fuel Properties of Eleven Vegetable Oils. *Transactions of the ASAE* 25, no. 6: 1472-1483.
7. Labeckas, G., and S. Slavinskas. 2006. Performance of direct-injection off-road diesel engine on rapeseed oil. *Renewable Energy* 31, no. 6: 849-863. dx.doi.org/10.1016/j.renene.2005.05.009.
8. Nwafor, O. M. I. 2001. Emission characteristics of neat rapeseed oil fuel in diesel engine. *International Journal of Ambient Energy* 22, no. 3 (July): 146-154.
9. Peterson, C. L., G. L. Wagner, and D. L. Auld. 1983. Vegetable Oil Substitutes for Diesel Fuel. *Transactions of the ASAE* 26, no. 2: 322-332.

10. Prasad, C. M. V., M. V. S. Krishna, C. P. Reddy, and K. R. Mohan. 2000. Performance evaluation of non-edible vegetable oils as substitute fuels in low heat rejection diesel engines. *Proceedings of the Institution of Mechanical Engineers, Part D: Journal of Automobile Engineering* 214, no. 2: 181-187. dx.doi.org/10.1243/0954407001527330.
11. Pryor, R. W., M. A. Hanna, J. L. Schinstock, and L. L. Bashford. 1983. Soybean Oil Fuel in a Small Diesel Engine. *Transactions of the ASAE* 26, no. 2: 333-337.
12. Pugazhvadivu, M., and K. Jeyachandran. 2004. Effect of fuel injection pressure and preheating on the performance and emissions of a vegetable oil fuelled diesel engine. *Proceedings of the Fall Technical Conference of the ASME International Combustion Engine Division, Oct 24-27 2004*: 479-486.
13. Radu, R., E. Rakosi, C. Iulian-Agape, and R. Gaiginschi. 2001. Application of a combustion model to a Diesel engine fueled with vegetable oils. *JSME International Journal, Series B: Fluids and Thermal Engineering* 44, no. 4: 634-640. dx.doi.org/10.1299/jsmeb.44.634.
14. Raja, A.S., G. Lakshmi Narayana Rao, N. Nallusamy, and M. Selva Ganesh Kumar. 2004. Effect of combustion chamber design on the performance and emission characteristics of a direct injection (Dl) diesel engine fuelled with refined rice bran oil blends. *Proceedings of the Fall Technical Conference of the ASME International Combustion Engine Division, Oct 24-27 2004*: 427-435.
15. Ryan, T.W., L.G. Dodge, and T.J. Callahan. 1984. The effects of vegetable oil properties on injection and combustion in two different diesel engines. *Journal of the American Oil Chemists' Society* 61, no. 10 (October 5): 1610-1619. dx.doi.org/10.1007/BF02541645.
16. Srivastava, A., and R. Prasad. 2000. Triglycerides-based diesel fuels. *Renewable and Sustainable Energy Reviews* 4, no. 2: 111-133. dx.doi.org/10.1016/S1364-0321(99)00013-1.
17. Yoshimoto, Y. 2003. Performance and emissions of diesel fuels containing rapeseed oil and the characteristics of evaporation and combustion of single droplets. *SAE Technical Paper Series*: 2003-01-3201.
18. web.archive.org/web/20070120171233/eere.energy.gov/afdc/pdfs/afv_info.pdf
19. ASTM D975.
20. ASTM 6751.

21. Bandel, W. and W. Heinrich, "Vegetable Oil Derived Fuels and Problems Related to their Use in Diesel Engines," in *Energy from Biomass* (New York, NY: Applied Science, 1982), 822.
22. Bari, S. 2004a. Investigation into the deteriorated performance of diesel engine after prolonged use of vegetable oil. *Proceedings of the Fall Technical Conference of the ASME International Combustion Engine Division, Oct 24-27 2004*: 447-455.
23. ———. 2002b. Performance deterioration and durability issues while running a diesel engine with crude palm oil. *Proceedings of the Institution of Mechanical Engineers, Part D: Journal of Automobile Engineering* 216, no. 9: 785-792. dx.doi.org/10.1243/09544070260340871.
24. Peterson, C. L., G. L. Wagner, and D. L. Auld. 1983. Vegetable Oil Substitutes for Diesel Fuel. *Transactions of the ASAE* 26, no. 2: 322-332.
25. Pryor, R. W., M. A. Hanna, J. L. Schinstock, and L. L. Bashford. 1983. Soybean Oil Fuel in a Small Diesel Engine. *Transactions of the ASAE* 26, no. 2: 333-337.
26. Adams, C., J.F. Peters, M.C. Rand, B.J. Schroer, and M.C. Ziemke. 1983. Investigation of soybean oil as a diesel fuel extender: Endurance tests. *Journal of the American Oil Chemists' Society* 60, no. 8: 1574-1579. dx.doi.org/10.1007/BF02666588.
27. Ziejewski, M., and K. R. Kaufman. 1983. Laboratory endurance test of a sunflower oil blend in a diesel engine. *Journal of the American Oil Chemists' Society* 60, no. 8: 1567-1573. dx.doi.org/10.1007/BF02666587.
28. Borgelt, S.C., and F.D. Harris, "Endurance Tests Using Soybean Oil-Diesel Fuel Mixture to Fuel Small Pre-Combustion Chamber Engines," in *Vegetable Oil Fuels: Proceedings of the International Conference on Plant and Vegetable Oils as Fuels* (Fargo, ND, USA: ASAE, St. Joseph, MI, USA, 1982), 364-373.
29. Fort, E.F., and P.N. Blumberg, "Performance and Durability of a Turbocharged Diesel Fueled with Cottonseed Oil Blends," in *Vegetable Oil Fuels: Proceedings of the International Conference on Plant and Vegetable Oils as Fuels* (Fargo, ND, USA: ASAE, St. Joseph, MI, USA, 1982), 374-383.
30. Ryan, T.W., T. J., Callahan, and L. G. Dodge, "Characterization of Vegetable Oils for Use as Fuels in Diesel Engines," in *Vegetable Oil Fuels:*

Proceedings of the International Conference on Plant and Vegetable Oils as Fuels (St. Joseph, MI: ASAE, 1982), 70-81.

30a: A. Ryan T. W. , L. G. Dodge, and T. J. Callahan, "Effects of Vegetable Oil Properties on Injection and Combustion in Two Different Diesel Engines," *Journal of the American Oil Chemists' Society* 61, no. 10 (1983): 1615.

B. Acaroglu, H. Oguz, and H. Ogut, "An investigation of the use of rapeseed oil in agricultural tractors as engine oil," *Energy Sources* 23, no. 9 (2001): 826, dx.doi.org/10.1080/009083101316931898.

C. Lang W., S. Sokhansanj, and F. W. Sosulski, "Modelling the Temperature Dependence of Kinematic Viscosity for Refined Canola Oil," *Journal of the American Oil Chemist's Society* 69, no. 10 (1992): 1055.

D. Silvio C. A. De Almeida, et al., "Performance of a diesel generator fuelled with palm oil," *Fuel 81*, no. 16 (2002): 2098, dx.doi.org/10.1016/S0016-2361(02)00155-2.

E. Pugazhvadivu M., and K. Jeyachandran, "Investigations on the performance and exhaust emissions of a diesel engine using preheated waste frying oil as fuel," *Renewable Energy* 30, no. 14 (2005): 2193, dx.doi.org/10.1016/j.renene.2005.02.001.

F. Karaosmanoglu F., et al., "Fuel properties of cottonseed oil," *Energy Sources* 21, no. 9 (1999): 821-828, dx.doi.org/10.1080/00908319950014371.

G. 1. Altin R., S. Cetinkaya, and H. S. Yucesu, "Potential of using vegetable oil fuels as fuel for diesel engines," *Energy Conversion and Management* 42, no. 5 (2001): 532, dx.doi.org/10.1016/S0196-8904(00)00080-7.

H. Labeckas, G., and S. Slavinskas, "Performance of direct-injection off-road diesel engine on rapeseed oil," *Renewable Energy* 31, no. 6 (2006): 849,852, dx.doi.org/10.1016/j.renene.2005.05.009.

I. Pugazhvadivu M., and K. Jeyachandran, "Effect of fuel injection pressure and preheating on the performance and emissions of a vegetable oil fuelled diesel engine," *Proceedings of ICEF04*, 2004: 4

31. Pugazhvadivu, M., and K. Jeyachandran, "Investigations on the performance and exhaust emissions of a diesel engine using preheated

waste frying oil as fuel," *Renewable Energy* 30, no. 14 (2005): 2189-2202, dx.doi.org/10.1016/j.renene.2005.02.001.

32. Nwafor, O. M. I., "The effect of elevated fuel inlet temperature on performance of diesel engine running on neat vegetable oil at constant speed conditions," *Renewable Energy* 28, no. 2 (2002): 171-181, dx.doi.org/10.1016/S0960-1481(02)00032-0.
33. Nwafor, O. M. I.,"Emission characteristics of diesel engine running on vegetable oil with elevated fuel inlet temperature," *Biomass and Bioenergy* 27, no. 5 (2004): 507-511, dx.doi.org/10.1016/j.biombioe.2004.02.004.
34. Kumar, M., Senthil A., Kerihuel, J. Bellettre, and M. Tazerout, "Experimental investigations on the use of preheated animal fat as fuel in a compression ignition engine," *Renewable Energy* 30, no. 9 (July 2005): 1443-1456, sciencedirect.com/science/article/B6V4S-4F1SVD3-5/2/512c244e67c49b4a2a2b6e2cbdb1ed53.
35. Bari, S. C., W. Yu, and T. H. Lim,"Performance deterioration and durability issues while running a diesel engine with crude palm oil," Proceedings of the Institution of Mechanical Engineers, Part D: Journal of Automobile Engineering 216, no. 9 (2002): 785-792, dx.doi.org/10.1243/09544070260340871.
36. Santos, J. C., O. I. M. G. Santos, and A. G. Souza,"Effect of heating and cooling on rheological parameters of edible vegetable oils," *Journal of Food Engineering* 67, no. 4 (2005): 401-405, dx.doi.org/10.1016/j.jfoodeng.2004.05.007.
37. Goering, C. E. et al., "Fuel Properties of Eleven Vegetable Oils," Transactions of the ASAE 25, no. 6 (1982): 1472-1483.
38. Labeckas, G., and S. Slavinskas. 2006. Performance of direct-injection off-road diesel engine on rapeseed oil. *Renewable Energy* 31, no. 6: 849-863. dx.doi.org/10.1016/j.renene.2005.05.009.
39. Ryan, T.W., L.G. Dodge, and T.J. Callahan,"The effects of vegetable oil properties on injection and combustion in two different diesel engines," *Journal of the American Oil Chemists' Society* 61, no. 10 (October 5, 1984): 1610-1619, dx.doi.org/10.1007/BF02541645 (accessed July 20, 2007).
40. ASTM D613, in the U.S.A.

40a. Callahan, T.J., T.W. Ryan III, L.G. Dodge, and J.A. Schwald. 1988. "Effect of Fuel Properties on Diesel Spray Characteristics." *SAT Technical Paper Series*: 870533.

40b. Ibid.
41. Ryan, T.W., L.G. Dodge, and T.J. Callahan, "The effects of vegetable oil properties on injection and combustion in two different diesel engines," *Journal of the American Oil Chemists' Society* 61, no. 10 (October 5, 1984): 1610-1619, dx.doi.org/10.1007/BF02541645 (accessed July 20, 2007).
42. Goering, C.E., et al., "Fuel Properties of Eleven Vegetable Oils." *Transactions of the ASAE* 25, no. 6: 1472-1483.
43. Rakopoulos, C.D., K.A. Antonopoulos, and D.C. Rakopoulos, "Multi-zone modeling of Diesel engine fuel spray development with vegetable oil, bio-diesel or Diesel fuels," *Energy Conversion and Management* 47, no. 11-12 (July 2006): 1550-1573, (accessed September 24, 2007).
44. Fernando, S., M. Hanna, and S. Adhikari, "Lubricity characteristics of selected vegetable oils, animal fats, and their derivatives," *Applied Engineering in Agriculture* 23, no. 1 (2007): 5-11, asae.frymulti.com.proxy.lib.utk.edu:90/request.asp?JID=3&AID=22324&CID=aeaj2007&v=23&i=1&T=1.
45. Wypych, G., "Kerosene," in *Knovel Solvents: A Properties Database* (ChemTec Publishing, 2000), knovel.com/knovel2/Toc.jsp?BookID=635&VerticalID=0. (Kerosene is a slightly better solvent than petrodiesel)
46. King, J.W., "Determination of the Solubility Parameter of Soybean Oil by Inverse Gas-Chromatography," *Lebensmittel Wissenschaft & Technologie* 28, no. 2 (1995): 190-195.
47. Hu, J., et al., "Study on the solvent power of a new green solvent: Biodiesel," *Industrial and Engineering Chemistry Research* 43, no. 24 (2004): 7928-7931, dx.doi.org/10.1021/ie0493816 (accessed July 22, 2007).
48. Burke, J. "Solubility Parameters: Part 3 - Other Practical Solubility Scales," *Solubility Parameters: Theory and Application*, palimpsest.stanford.edu/byauth/burke/solpar/solpar3.html (accessed September 21, 2007).
49. Wijayasinghe and Makey, "Cooking Oil: A Home Fire Hazard in Alberta, Canada," *Fire Technology* 33, no. 2 (May 1, 1997): 140-166, dx.doi.org/10.1023/A:1015395001403 (accessed August 2, 2007).
50. Ibid.
51. Ibid.

52. Diesel fuel and exhaust emissions. World Health Organization, 1996. inchem.org/documents/ehc/ehc/ehc171.htm.
53. Rakopoulos, C.D., K.A. Antonopoulos, and D.C. Rakopoulos, "Multi-zone modeling of Diesel engine fuel spray development with vegetable oil, bio-diesel or Diesel fuels," *Energy Conversion and Management* 47, no. 11-12 (July 2006): 1550-1573, (accessed September 24, 2007).
54. List, G.R., T. Wang, and V.K.S. Shukka, "Storage, Handling, and Transport of Oils and Fats," in *Bailey's Industrial Oil and Fat Products*, Volume 5, ed. F. Shahidi (John Wiley & Sons, 2005), 191-229, knovel.com/knovel2/Toc.jsp?BookID=1432&VerticalID=0.
55. Wheeler, "Peroxide formation as a measure of autoxidative deterioration," *Journal of the American Oil Chemists' Society* 9, no. 4 (April 8, 1932): 89-97, dx.doi.org/10.1007/BF02553782 (accessed August 2, 2007).
56. The instability of this hydrogen-carbon bond is due to delocalization of the carbon-carbon double bond.
57. The best I can do is to point you to two sources: Schaich, K.M., "Storage, Handling, and Transport of Oils and Fats," in *Bailey's Industrial Oil and Fat Products*, Volume 1, ed. F. Shahidi (John Wiley & Sons, 2005), 269-355, knovel.com/knovel2/Toc.jsp?BookID=1432&VerticalID=0. E.N. Frankel, *Lipid Oxidation* (Bridgewater: P.J. Barnes & Associates, 2005). These two sources probably represent the edge of knowledge on the subject, and neither comes near to completely characterizing the processes of oxidation decomposition and polymerization. Such is the current state of knowledge.
58. Gupta, M.K., "Frying Oils," in *Bailey's Industrial Oil and Fat Products*, ed. F. Shahidi (John Wiley & Sons, 2005), 1-31, media.wiley.com.proxy.uchicago.edu/product_data/excerpt/92/04713854/0471385492.pdf
59. Alovert, M., *Biodiesel Homebrew Guide* (2005).
60. Gupta, M.K., "Frying Oils," in *Bailey's Industrial Oil and Fat Products*, ed. F. Shahidi (John Wiley & Sons, 2005), 1-31, media.wiley.com.proxy.uchicago.edu/product_data/excerpt/92/04713854/0471385492.pdf
61. Li, X., J. Li, and C. Sun, "Properties of Transgenic Rapeseed Oil based Dielectric Liquid," SoutheastCon, 2006. Proceedings of the IEEE, 2006, dx.doi.org/10.1109/second.2006.1629328.

62. solpower.com/soltron/soltronmain.asp
63. This inability of the standard-specified test to measure emulsified water is acceptable for diesel, biodiesel, and even unused cooking oil, because water is not easily emulsified in these fuels. Used cooking oil, on the other hand, will almost certainly contain significant emulsified water because of the creation and introduction of emulsifiers during cooking.

Chapter 3

1. A micron is .001 millimeter or .00003937 inch. Most human hairs are between 55 and 115 microns in diameter (Hardy, Daniel."Quantitative Hair Form Variation in Seven Populations". *American Journal of Physical Anthropology* (39:1, 1973), 7-17.

Chapter 4

1. The tractor had a Valmet 311DS engine, a direct injection, turbocharged, three- cylinder engine.
2. The Iodine Value of an oil goes down as an oil oxidizes.
3. Sunwizard is the online alias of the engineer that has designed this system and described it on an online forum dedicated to discussing vegetable oil fuel. He has asked to be referred to by that alias in this book.
4. According to Sunwizard and Dieselcraft, this increase in 6-10 micron particles is likely due to sampling error, and is not reflective of the actual effects of the centrifuge.
5. sandybrae.com/
6. noria.com/learning_center/category_article.asp?articleid=301
7. Title 42 U.S. Code, Sec 7522. 2005ed. law.cornell.edu/uscode/html/uscode42/usc_sec_42_00007522——000-.html
8. ibid.
9. Dear Manufacturer's Guidance Letter CIS-06-02, epa.gov/otaq/cert/dearmfr/cisd0602.pdf
10. ibid.

Appendix B

1. Ryan, T.W., L.G. Dodge, and T.J. Callahan,"The effects of vegetable oil properties on injection and combustion in two different diesel engines,"

Journal of the American Oil Chemists' Society 61, no. 10 (October 5, 1984): 1610-1619, dx.doi.org/10.1007/BF02541645 (accessed July 20, 2007).
2. Remmele, E., "Pre-Standard DIN V 51605 for Rapeseed Oil Fuel," *15th European Biomass Conference & Exhibition*, May 2007.
3. Pryde, E. H., "Vegetable Oil Fuel Standards," *Vegetable Oil Fuels*, August 1982.

Appendix C

1. dieselpowermag.com/tech/dodge/0604dp_cummins_diesel_motor_history/index.html
2. musclemustangfastfords.com/tech/mmfs_070025_power_stroke_diesel_history/index.html
3. vegistroke.com

Index

Page locators in **bold** indicate tables and page locators in *italics* indicate figures.

About the Author

Forest Gregg has worked as a researcher with Frybrid — a recognized leader in the technical development of vegetable oil conversion systems — and as a designer, fabricator, and installer of conversion systems. His first experience with using vegetable oil as a diesel fuel was after he started a small circus that travelled the country in an old New York City schoolbus that was converted, badly, to burn vegetable oil. After that educational experience, he decided to learn how to do it right. Forest now resides in the beautiful mountains of North Carolina, and tries to keep his fingernails clean, but fails.

New Society Publishers

ENVIRONMENTAL BENEFITS STATEMENT

New Society Publishers has chosen to produce this book on Enviro 100, recycled paper made with **100% post consumer waste**, processed chlorine free, and old growth free.

For every 5,000 books printed, New Society saves the following resources:[1]

24	Trees
2,188	Pounds of Solid Waste
2,408	Gallons of Water
3,140	Kilowatt Hours of Electricity
3,978	Pounds of Greenhouse Gases
17	Pounds of HAPs, VOCs, and AOX Combined
6	Cubic Yards of Landfill Space

[1]Environmental benefits are calculated based on research done by the Environmental Defense Fund and other members of the Paper Task Force who study the environmental impacts of the paper industry.

For more information on this environmental benefits statement, or to inquire about environmentally friendly papers, please contact New Leaf Paper – info@newleafpaper.com Tel: 888 • 989 • 5323.

NEW SOCIETY PUBLISHERS